RETHINKING CONSCIOUSNESS

ALSO BY MICHAEL GRAZIANO

Consciousness and the Social Brain

*The Spaces Between Us: A Story of Neuroscience,
Evolution, and Human Nature*

The Intelligent Movement Machine

*God, Soul, Mind, Brain: A Neuroscientist's Reflections
on the Spirit World*

MICHAEL S. A. GRAZIANO

RETHINKING
CONSCIOUSNESS

A Scientific Theory of
Subjective Experience

W. W. NORTON & COMPANY
Independent Publishers Since 1923

For information about permission to reproduce selections from this book, write to
Permissions, W. W. Norton & Company, Inc., 500 Fifth Avenue, New York, NY 10110

For information about special discounts for bulk purchases, please contact
W. W. Norton Special Sales at specialsales@wwnorton.com or 800-233-4830

Manufacturing by Sheridan
Book design by Ellen Cipriano
Production manager: Lauren Abbate

Library of Congress Cataloging-in-Publication Data

Names: Graziano, Michael S. A., 1967– author.
Title: Rethinking consciousness : a scientific theory of
subjective experience / Michael S.A. Graziano.
Description: First Edition. | New York : W.W. Norton & Company, 2019. |
Includes bibliographical references and index.
Identifiers: LCCN 2019014794 | ISBN 9780393652611 (hardcover)
Subjects: LCSH: Consciousness. | Experience—Psychological aspects.
Classification: LCC B808.9 .G73 2019 | DDC 153—dc23
LC record available at https://lccn.loc.gov/2019014794

W. W. Norton & Company, Inc., 500 Fifth Avenue, New York, N.Y. 10110
www.wwnorton.com

W. W. Norton & Company Ltd., 15 Carlisle Street, London W1D 3BS

1 2 3 4 5 6 7 8 9 0

For Ben and Ele

CONTENTS

＊

RETHINKING
CONSCIOUSNESS

CHAPTER 1

The Elephant in the Room

WHEN MY SON was 3 years old, I made his favorite stuffed elephant talk. At that age he couldn't tell how bad a ventriloquist I was, so the trick worked very well on him. He loved it. Over the next several years, as I improved my technique, the uncanny power of that illusion began to impress me. Ventriloquism is not just a voice that comes out of a puppet, as though out of a hidden speaker. Even in the hands of a mediocre performer like me, something special happens. The puppet comes to life with its own personality, and consciousness seems to emanate from it.

The human brain clearly must contain machinery that impels us to attribute consciousness to the puppet. But we didn't evolve that machinery to enjoy ventriloquism. Humans are social animals, and we routinely use the same trick on each other. When I talk to someone, I have an automatic impression of thoughts, emotions, and awareness emanating from that person. I'm not directly perceiving the person's mind, of course. Instead, my brain is constructing a handy model of a mind and projecting it onto the person, treating that person like my son treats the puppet.

We apply the same process to more than just people. We attribute awareness to our pet cats and dogs, and some people even swear that their houseplants are conscious. The ancients felt sure that trees and rivers were sentient; children perceive consciousness in their favorite toys; and heck, the other day I got mad at my computer. So I'm not talking about intellectually figuring out whether something has a mind or cleverly deducing what might be in that mind—although we do that, too. I'm talking about an automatic, gut intuition, which is often wrong but sometimes persuasively potent, that an essence of awareness is emanating from an object.

As I thought more about ventriloquism, I began to wonder if my own consciousness and these examples of attributing consciousness to others might stem from the same source. Maybe there is one unifying explanation: we automatically build models of minds and project them onto ourselves and other people. Our intuitions about a mysterious conscious presence, our conviction that it is present in me or you or this pet or that object, might depend on those simplified but useful models—sets of information that the brain constructs to understand its world.

This is the kind of profound insight that can only come from talking to a stuffed elephant. It also diverted my scientific work to the study of consciousness.

For 20 years, I had been studying more traditional issues in neuroscience—how the brain monitors the space immediately around the body and how it controls complex movements within that space.[1] That background in basic, nuts-and-bolts neuroscience turned out to be useful for building a theory of consciousness. In 2010, my colleagues and I began outlining what we called the attention schema theory, drawing on data from neuroscience, psychology, and evolution and adding insights from engineering.[2] The theory is part of a larger change of perspective in the scientific community.[3] The new approach does not solve the so-called "hard problem" of consciousness—how a physical brain can generate a nonphysical essence.[4] Instead, it explains why people might mistakenly think that there is a hard problem to begin with, why that mistaken intuition is built deep into us where we're unlikely to change it, and why its presence is advantageous, maybe even necessary, for the functioning of the brain.

I first understood the theory from the point of view of social interaction. At its root, however, the theory depends on a more general property of the brain: model-based knowledge.[5] The brain constructs internal models—ever-changing rich packets of information, constructed continuously and automatically, like bubbles of meaning that lie beneath the level of higher thought or of language. Those internal models represent important items that are useful to monitor, sometimes external objects and sometimes aspects of the self. The representations are simplified and distorted, like impressionistic or cubist paintings of reality, and we report the content of them as though we are reporting literal reality. We can't help it—they come built into us. Our intuitive understanding of the world around us and our understanding of ourselves, always distorted and simplified, are dependent on those internal models.

In the theory, our metaphysical intuitions about ourselves, about consciousness as a nonphysical inner essence—sometimes called the "ghost in the machine"[6]—are derived from a particular internal model. I call it the attention schema, for reasons that will become clear throughout this book. It is a simplified depiction of how the brain seizes on information and deeply processes it. That depiction is an efficient way for the brain to understand and monitor its own internal abilities. The same kind of internal model can also be used, to a lesser extent, to monitor and make predictions about other people.

This model-based approach can sometimes sound like a dismissal or a devaluing of consciousness—but it is decidedly not. The internal model that tells us we are conscious is deep, rich, continuous, and probably necessary. Almost nothing we do—perceiving, thinking, acting, socially interacting—would work properly without that part of the system.

IN THIS BOOK, I will use the terms *consciousness*, *subjective awareness*, and *subjective experience* interchangeably, although I acknowledge that those words are not always used by scholars in an equivalent way. The word *consciousness* is especially notorious for its many slippery connotations. I want to clarify, first, what I *don't* mean by it, before I get to what I do mean. Sometimes people think of consciousness as the ability to know who you are and to understand your trajectory through life. Other people think of it more as the ability to process the world around you and on that basis to make intelligent decisions. I mean neither of those things.

The most common understanding of the internal experience is probably the "stream of consciousness," the constantly changing, kaleidoscopic contents of the mind. It's that riot of stuff in your head that James Joyce famously captured in his 1922 novel *Ulysses*.[7] Joyce meticulously recorded the ever-changing sight and sound and touch of the world, the taste and smell, the memories of the recent and distant past boiling up, a running internal dialogue, the conflicting emotions and fantasies, some of them so scandalous that the book was initially banned. (The 1933 court case, *The United States v. One Book Called Ulysses*, gave us our modern legal definition of obscenity.) But again, this is not what I mean by consciousness. That stream of material is not very well defined, and its sheer volume is overwhelming to study scientifically.

Instead, imagine putting 10,000 odds and ends in a bucket. You can catalog that complicated list of items, as James Joyce did. But you can also ask a more basic question: what about the bucket? Never mind the contents for now. What is the bucket made of, and where did it come from? How does a person get to be conscious of anything at all? Consciousness can't be just the information inside us, because we're conscious of only a small amount of the huge pool of information in the brain at any one time. Something must happen to a limited amount of information to make us conscious of it. What makes that happen? That more specific question has increasingly occupied philosophers and scientists.[8] The term *consciousness* has come to mean the act of being conscious of something, rather than the material of which you are conscious.

I suspect that the gradual shift in philosophy from focusing on the many items in a stream of consciousness to the act of *being conscious* has something to do with the advance of computer technology over the past half-century. As our information technology has improved, the information content of the mind has become less mysterious, while at the same time the act of being conscious of it, of experiencing anything at all, has become more remote and seemingly unsolvable. Let's look at a few examples.

You can connect a digital camera to a computer and program the system to process the incoming visual information. The computer can extract color, shape, and size, and it can identify objects. The human brain does something similar. The difference is that people also have a subjective experience of what they see. We don't just register the information that the object is red; we have an *experience* of redness. Seeing *feels* like something. A modern computer can process a visual image, but engineers have not yet solved how to make the computer conscious of that information.

Now consider something a little more personal than visual perception: the autobiographical memories that define your trajectory through life. Nothing typifies the Joycean stream of consciousness so much as the memories that are constantly bubbling up. And yet we know how to build a machine that stores and retrieves memory. Every computer has the capacity, and scientists know the general principles, if not the details, of how memory is stored in the brain. Memory is not a fundamental mystery. It also doesn't cause consciousness. Again, the stuff *in* consciousness—in this example, a memory—is not the same as the act of *being conscious* of a memory.

I'll give one more example: decision-making. If anything defines the mystery of human consciousness, surely it must be our ability to make decisions. We take in information, process it, judge it, and make a choice about what to do next. But, again, I would say that consciousness is not an intrinsic part of decision-making. All computers make decisions. In a sense, that's the definition of a computer. It takes in information, manipulates it, and uses it to select one course of action out of many. Most of the decisions made by the human brain, possibly tens of thousands a day, occur automatically with no subjective experience. In a few select instances, we report a subjective awareness of making a decision. Sometimes we call it intention, choice, or free will. But the mere ability to make a decision does not require consciousness.

With these and many other examples, the rise of computer technology has revealed the distinction between the content of consciousness, which is increasingly well understood at an engineering level, and the act of being conscious of it. My interest lies in that crucial, second part of the puzzle: how do we get to have a subjective experience of anything at all?

Sometimes people find that focus limiting. I've often been asked: What about memory? What about conscious choice? What about self-understanding? What about intentions and beliefs? Aren't these things the bread and butter of consciousness? I agree: all of these are important concerns and are crucial objects in the bucket of human consciousness. Still, they are not fundamental mysteries. They are matters of information processing, and we can imagine, at least in principle, how to engineer them. The fundamental mystery is the bucket itself. What is consciousness—what is it made of? How can something enter it, what is gained by entering it, and why do so few items in the brain end up there?

Traditionally, scholars assumed that something as amorphous and slippery as consciousness must be impossible to understand scientifically. But given recent insights, I am now pretty sure that it is as understandable and buildable as visual processing, memory, decision-making, or any other specific item that makes up its content.

I'VE WRITTEN ABOUT consciousness many times before. This book, however, is written entirely for the general reader. In it, I attempt to spell out, as simply and clearly as possible, a promising scientific theory of consciousness—one that can apply equally to biological brains and artificial machines.

The next few chapters start with evolution. Beginning more than half a billion years ago with the appearance of neurons, the cells that make up the brain, I'll describe the evolving complexity of the nervous system. Along the way I'll introduce the components of the attention schema theory, and by Chapter 6, the main scaffold of the theory will be in place.

I'll then discuss how the attention schema theory makes contact with other theories. The attention schema theory is one of about half a dozen main theories of consciousness that are gaining ground in the scientific literature. My impression, and the impression I try to convey in this book, is that these theories should not always be viewed as rivals, and we should not wait to see which one kills off its competitors. As different as they are—and I do disagree with a lot of what has been proposed—these many theories can also have strange, hidden connections to each other. Each one contributes important insights. I believe we are beginning to see the glimmerings of a consensus view—or maybe more like a consensus web of ideas.

The final chapters take a deep dive into the technological implications. We are close to understanding consciousness well enough to build it, and when we do, the new technology is likely to transform our civilization. Machine consciousness is just the first step. If consciousness is engineerable, then the mind is, in principle, migratable from one device to another. Though much farther down the road, it is theoretically possible to read the relevant data from a human brain and migrate that person's mind to an artificial platform.[9] The technology could allow minds to live indefinitely and to explore environments, such as interstellar space, that are hostile to biological bodies. No laws of physics stand in the way—only gadgetry that has yet to be invented.

If consciousness can be understood from a scientific and engineering perspective, then the topic is no longer just a philosophical game for scholars. It becomes an urgent practical matter. This book will follow the uses of consciousness to many possible technological futures, some good and some admittedly horrible. But good or bad, I am now pretty sure that a scientific understanding of consciousness and an ability to engineer it artificially are rapidly approaching.

CHAPTER 2

Crabs and Octopuses

SELF-REPLICATING, BACTERIAL LIFE first appeared on Earth about 4 billion years ago. For most of Earth's history, life remained at the single-celled level, and nothing like a nervous system existed until around 600 or 700 million years ago (MYA). In the attention schema theory, consciousness depends on the nervous system processing information in a specific way. The key to the theory, and I suspect the key to any advanced intelligence, is attention—the ability of the brain to focus its limited resources on a restricted piece of the world at any one time in order to process it in greater depth. In this and the next several chapters, I'll examine how attention may have evolved from early animals to humans and how the property we call consciousness may have emerged along with it.[1]

I will begin the story with sea sponges, because they help to bracket the evolution of the nervous system. They are the most primitive of all multicellular animals, with no overall body plan, no limbs, no muscles, and no need for nerves. They sit at the bottom of the ocean, filtering nutrients like a sieve. And yet sponges do share some genes with us, including at least 25 that, in people, help structure the nervous system.[2] In sponges, the same genes may be involved in simpler aspects of how cells communicate with each other. Sponges seem to be poised right at the evolutionary threshold of

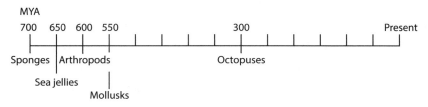

Figure 2.1 Invertebrates discussed in this chapter and
their approximate time of first appearance.

the nervous system. They are thought to have shared a last common ancestor
with us between about 700 and 600 MYA (see the time line in Figure 2.1).[3]

In contrast, another ancient type of animal, the sea jelly, does have a
nervous system. Sea jellies don't fossilize very well, but by analyzing their
genetic relationship to other animals, biologists estimate that they may
have split from the rest of the animal kingdom as early as 650 MYA.[4]
These numbers may change with new data, but as a plausible, rough
estimate, it seems that neurons, the basic cellular components of a nervous system, first appeared in the animal kingdom somewhere between
sponges and sea jellies, a little more than half a billion years ago.

A NEURON IS, in essence, a cell that transmits a signal. A wave of electrochemical energy sweeps across the membrane of the cell from one end
to the other, at about 200 feet per second, and influences another neuron,
a muscle, or a gland. The earliest nervous systems may have been simple
nets of neurons laced throughout the body, interconnecting the muscles.
Hydras work on this nerve-net principle.[5] They are tiny water creatures—
transparent, flowerlike animals with sacs for bodies attached to many
arms—and belong to the same ancient category as sea jellies. If you touch
a hydra in one place, the nerve net spreads the signals indiscriminately,
and the hydra twitches as a whole.

A nerve net doesn't process information—not in any meaningful
sense. It merely transmits signals around the body. It connects the sensory

stimulus (a poke on the hydra) to a muscle output (a twitch). After the emergence of the nerve net, however, nervous systems rapidly evolved a second level of complexity: the ability to enhance some signals over others. This simple but powerful trick of signal boosting is one of the basic ways that neurons manipulate information. It is a building block of almost all computations that we know about in the brain.

The eye of the crab is one of the best-studied examples.[6] The crab has a compound eye with an array of detectors, each with a neuron inside it. If light falls on one detector, it activates the neuron inside. So far so good. But in an added pinch of complexity, each neuron is connected to its nearest neighbors, and because of those connections, the neurons compete with each other. When a neuron in one detector becomes active, it tends to suppress the activity of the neurons in the neighboring detectors, like a person in a crowd who is trying to shout the loudest while shushing the people nearest to him.

The result is that if a blurry spot of light shines on the crab's eye, with the brightest part of the spot hitting one detector, the neuron in that detector becomes highly active, wins the competition, and shuts down its neighbors. The pattern of activity across the set of detectors in the eye not only signals a bright spot, but also signals a ring of darkness around it. The signal is, in this way, enhanced. The crab eye takes a fuzzy, grayscale reality and sharpens it into a high-contrast image with exaggerated, brighter peaks and darker shadows. This signal enhancement is a direct consequence of neurons inhibiting their neighbors, a process called *lateral inhibition*.[7]

The mechanism in the eye of a crab is arguably the simplest and most fundamental example—the model A case—of attention. Signals compete with each other, the winning signals are boosted at the expense of the losing signals, and those winning signals can then go on to influence the animal's movements. *That* is the computational essence of attention. Our human attention is merely an elaborated version of it, made of the same building blocks. You can find the crab-eye method of lateral inhibition at every stage of processing in the human nervous system, from the eye to the highest levels of thought in the cerebral cortex. The origin of attention lies deep in evolutionary time, more than half a billion years ago, with a surprisingly simple innovation.

Crabs belong to an extensive group of animals, the arthropods, which includes spiders and insects and other creatures with hard, jointed exoskeletons and which branched off from other animals about 600 MYA.[8] The most famous extinct arthropod, the one with the biggest fan club today, is the trilobite—a leggy, jointed creature almost like a miniature horseshoe crab, which crawled about the bottom of Cambrian seas as early as 540 MYA. When trilobites died and sank into very fine silt on the ocean floor, their faceted eyes were sometimes fossilized in amazing detail.[9] If you look at a trilobite fossil and examine its bulging eyes through a magnifying glass, you can often still see the orderly mosaic of individual detectors. Judging from these fossilized details, the trilobite's eye must have closely resembled a modern crab's eye in its organization and is likely to have used the same trick of competition between neighboring detectors to sharpen its view of the ancient seabed.

IMAGINE AN ANIMAL built piecemeal with "local" attention. In that animal, each part of the body would function like a separate device, filtering its own information and picking out the most salient signals. One of the eyes might say, "This particular spot is especially bright. Never mind the other spots." Meanwhile, independently, one of the legs says, "I've just been poked hard right here. Ignore the lighter touches nearby!" An animal with only this capability would act like a collection of separate agents that happen to be physically glued together, each agent shouting out its own signals, triggering its own actions. The animal's behavior would be, at best, chaotic.

For a coherent response to its environment, the animal needs a more centralized attention. Can many separate sources of input—the eyes, the body, the legs, the ears, the chemical sensors—pool their information together in one place for a global sorting and a competition among signals? That convergence would allow the animal to select the most vivid object in its environment, the one that seems most important at the moment, and then generate a single, meaningful response.

Nobody knows when that type of centralized attention first appeared, partly because nobody is certain which animals have it and which ones don't. Vertebrates have a central attention processor, which I'll describe in the next chapter. But the mechanisms of attention have not been as thoroughly studied in invertebrates. Many types of animals, such as segmented worms and slugs, do not have a central brain. Instead they have clusters of neurons, or ganglia, scattered throughout their bodies to perform local computations.[10] They probably don't have centralized attention.

Arthropods, such as crabs, insects, and spiders, are better candidates for centralized attention. They have a central brain, or at least an aggregate of neurons in the head that is larger than any of the others in their bodies.[11] That large ganglion may have evolved partly because of the requirements of vision. The eyes being in the head, and vision being the most complicated and information-intensive sense, the head gets the largest share of neurons. Some aspects of smell, taste, hearing, and touch also converge on that central ganglion. Insects are brainier than people think. When you swat at a fly and it manages to escape—as it almost always does—it isn't just darting away on a simple reflex. It probably has something that we can call central attention, or the ability to rapidly focus its processing resources on whatever part of its world is most important at the moment, in order to generate a coordinated response.[12]

OCTOPUSES ARE THE superstars of the invertebrates because of their astonishing intelligence. They're considered mollusks, like clams or snails. Mollusks probably first appeared about 550 MYA and remained relatively simple, at least in the organization of their nervous systems, for hundreds of millions of years.[13] One branch, the cephalopods, eventually evolved a complex brain and sophisticated behavior and may have reached something close to the modern form of an octopus around 300 MYA.[14]

Octopuses, squid, and cuttlefish are true aliens with respect to us.[15] No other intelligent animal is as far from us on the tree of life. They show us that big-brained smartness is not a one-off event, because it evolved independently at least twice—first among the vertebrates and then again among the invertebrates.

Octopuses are excellent visual predators. A good predator must be smarter and better coordinated than its prey, and using vision to locate and recognize prey is especially computationally intensive. No other sensory system has such a fire hose of varied information pouring in and such a need for an intelligent way to focus on useful subsets of that information. Attention, therefore, is the name of the game for a visual predator. Maybe that lifestyle has something to do with the expansion of octopus intelligence.

Whatever the reason, the octopus evolved an extraordinary nervous system. It can use tools, solve problems, and show unexpected creativity.[16] In a now classic demonstration, octopuses can learn to open a glass jar by unscrewing the top in order to get to a tasty morsel within. The octopus has a central brain and also an independent, smaller processor in each arm, giving it a unique mixture of centralized and distributed command.[17] The octopus also probably has self models—rich, constantly updated bundles of information to monitor its body and behavior. From an engineering perspective, it would need self models to function effectively. For example, it might have some form of a body schema that keeps track of the shape and structure of its body in order to coordinate movement. (Perhaps each arm has its own arm schema.) In that sense, you could say that an octopus knows about itself. It possesses information about itself and about the outside world, and that information results in complex behavior.

But all of these truly wonderful traits do not mean that an octopus is conscious.

Consciousness researchers sometimes use the term *objective awareness* to mean that the information has gotten in and is being processed in a manner that affects behavioral choice.[18] In that rather low-bar definition, one could say that a microwave is aware of the time setting and a self-driving car is aware of the looming obstacle. Yes, an octopus is objectively aware of itself and of the objects around it. It contains the information.

But is it *subjectively* aware? If it could talk, would it claim to have a subjective, conscious experience the same way that you or I do?

Let's ask the octopus. Imagine a somewhat improbable thought experiment—and remember the experiment, because it will come in handy throughout this book. Suppose we've gotten hold of a crazy science fiction device—let's call it the Speechinator 5000—that serves as an information-to-speech translator. It has a port that can be plugged into the octopus's head, and it verbalizes the information found in the brain.

It might say things like "There is a fish" if the octopus's visual system contains information about a nearby fish. The device might say, "I am an entity with a bunch of limbs that move in this and that way." It might say, "Getting a fish out of a jar requires turning that circular part." It would say many things, reflective of the information that we know is contained inside the octopus's nervous system. But we don't know if it would say, "I have a subjective, private experience—a consciousness—of that fish. I don't just process it. I *experience* it. Seeing a fish *feels* like something." We don't know if its brain contains that type of information because we don't know what the octopus's self models tell it. It may lack the machinery to model what consciousness is or to attribute that property to itself. Consciousness could be irrelevant to the animal.

The octopus conundrum is an instructive example of how an animal can be complex and intelligent, and yet we are, so far, unable to answer the question of its subjective experience or even whether the question has any meaning for that creature.

Maybe one source of confusion here is the automatic and powerful human urge to attribute consciousness to the objects around us. As I pointed out in Chapter 1, we are prone to see consciousness in puppets and other, even less likely objects. People sometimes believe that their house-plants are conscious. An octopus, with its richly complex behavior and its large eyes filled with focused attention, is a far more compelling inkblot test, so to speak, triggering a strong social perception in us. Not only do we know, intellectually, that it gathers objective information about its world, but we can't help feeling that it must have a subjective awareness as well emanating out of those soulful eyes. But the truth is, we don't know, and the sense we get of its conscious mind says more about *us* than about the octopus. The experts who study octopuses risk becoming the least reliable observers on this point, because they are the ones most likely to be entranced by these wonderful creatures. Later, in Chapter 5, I'll return to that pervasive human aspect of consciousness—how we use it as a tool in our social tool kit and reflexively attribute it to the agents around us.

Just to be clear, I'm not saying that octopuses are *not* conscious. But the octopus nervous system is still so incompletely understood that we can't yet compare its brain organization with ours and guess how similar it might be in its algorithms and self models. To make those types of comparisons, we will need to examine animals in our own lineage, the vertebrates.

CHAPTER 3

The Central Intelligence of a Frog

I GREW UP partly on a farm in upstate New York. Every summer, all night long, we'd hear a big bullfrog croaking out his mating song in the pond behind the house. We used to call him Elvis, and the smaller answering frog voice, Priscilla. I've been fond of frogs ever since, and when I became a neuroscientist, I was interested to read about what goes on inside their heads.

A frog has a part of the brain called the tectum. The word *tectum* means "roof" in Latin, and it's the largest, most obvious hump at the top of the brain. Frogs are not alone in this feature of the brain. The tectum may have been particularly studied in amphibians, but it is also present in fish, reptiles, birds, and mammals. All vertebrates have a tectum, but no other animals do, at least as far as we know. We can make a good guess that around half a billion years ago, a species of small, jawless fish, the common ancestor of vertebrates, evolved a tectum, and all its descendants inherited that brain part from it.[1]

People have a tectum, but ours is no longer at the top of the brain. It's a relatively tiny lump—or rather, two lumps, one on each side of the midline—buried beneath piles of brain structures that expanded in our evolutionary past. When found in people and other mammals, it's usually called the superior colliculus (which is Latin for "upper bump"). Here, for simplicity, I will call it a tectum.

For most of vertebrate evolution, the tectum was the pinnacle of intellectual achievement—the most complex, computationally sophisticated processor at the center of the brain. In a frog, the tectum takes in visual information and sorts the world into a literal map.[2] Each point on the rounded surface of the tectum corresponds to a point in space around the animal. The tectum on the right side of the frog's brain contains a neat, orderly map of the field of view of the left eye; and similarly for the left tectum and the right eye. When an erratic black dot zigzags around the frog, the eyes take in that information, the optic nerve sends the signals to the tectum, and the tectum triggers a set of muscle controllers. As a result, the tongue shoots out with impressive accuracy to snag the bug.

The logic of this input-output device was demonstrated particularly vividly by the neuroscientist Roger Sperry. In the 1960s, he performed surgery on a frog, removing the eyes, turning them upside down, and putting them back in.[3] The eyes took. Frogs have an amazing ability for regeneration. The optic nerve regrew from the eye to the tectum, reestablishing the internal visual map. When the frog was healed and could see again and a fly buzzed up above its head, its tongue shot down to the floor. If the fly buzzed to the frog's right, the tongue shot out to the left. The frog's central intelligence was a simple, beautifully efficient machine that collected specific nerve inputs and matched them with corresponding outputs. It was, unfortunately, tricked by scientific manipulation. The altered frog had to be fed by hand or it would have starved.

The frog's tectum is not limited to vision. It also collects information from the ears and from touch receptors across the skin.[4] A map of the frog's body surface, of the auditory space around the animal, and of visual space converge and are partly integrated in the tectum. It's the highest level of integration in the amphibian brain—the central processor that pulls together scattered signals pouring in from the environment, focuses on the most important event occurring at each moment in time, and then triggers a response.[5] The tectum is the frog's central attention mechanism.

SCIENTISTS CAN PROBE the brain with astonishing precision, like a computer engineer probing a circuit board. One standard method involves an electrode—a hair-thin, stiff wire coated in plastic insulation except at the tip. Only about a tenth of a millimeter of bare wire is exposed. Like a miniature detection wand, it can pick up electrical activity within a microscopically short distance of the bare metal. A long, flexible wire extends from the back end of the electrode, connecting it to a rack of equipment. The electrode is usually clamped in place by precision machinery and moved into position, 1 micrometer at a time, to study a targeted brain area.

The setup is sensitive enough to measure the activity of individual neurons in the brain. When a neuron near the electrode tip fires off a signal to its neighbors, the device picks up that tiny electrical pulse. The signal is amplified and piped to a loudspeaker, and the experimenter hears a click. Normally, a neuron might fire one or two clicks a second in a random pattern, but if an event occurs that recruits the neuron, the cell may fire off a sudden burst of clicks at a rate of 100 or so a second. One of the most thrilling pastimes of a neuroscientist is listening in on the clicking of an individual neuron and wondering what role it plays in the brain.

Each neuron in the frog's tectum acts like a detector.[6] The neuron monitors a particular region of space—for example, an area directly above the head—and its rate of clicking increases when an object enters that space. The neurons vary—some prefer a visual stimulus moving in a particular way, some prefer a touch or a sound. At least some of the neurons are multisensory. It makes no difference for a multisensory neuron if you move a visible object toward the top of the head, make a sound from that same upper location, or shut off the lights and touch the top of the head; the neuron clicks out its signal to the rest of the brain. If two or all three senses converge, carrying the same message about a nearby object, then the corresponding neurons in the tectum become especially active. A simple computation seems to be saying, "One piece of evidence, so far so good. Two or three converging pieces of evidence—something really important must be going on there!"[7]

The same basic experimental method can be used in the reverse direction, sending electrical pulses down the electrode to activate the nearby neurons—a method called microstimulation. The stimulation is so weak that you would not feel it on your skin, but it's enough to tickle neurons and induce them to generate their own signals. Microstimulation allows you to ask, If this small cluster of neurons near the electrode tip is artificially revved up, what does it tell the animal to do?

When a salamander's tectum is electrically stimulated, the animal pantomimes a complicated, coordinated movement.[8] It turns, opens its mouth, extends its tongue, stretches out its forelimbs, and shapes its long skinny fingers as if the animal were grabbing for prey. Whatever part of space is monitored by the neurons at a specific location in the tectum, when those neurons are electrically stimulated, they drive the animal to reach for that same location.

Stimulate a spot in the tectal map of an iguana, and the body, head, and eyes turn.[9] The animal looks directly at the part of space represented by that location in the map.

Stimulate the tectum of a fish, and it will turn its body to orient to the relevant part of space.[10] In that case, turning accurately toward a specific location is not as easy as swiveling a neck joint. It involves a complex interaction between the fins and the surrounding water.

Pit vipers, such as rattlesnakes or moccasins, have their own version of infrared vision: a pair of specialized, heat-sensing organs located about halfway between their eyes and nostrils. Those organs send information to the tectum, which contains a map of heat signals overlaid on the more usual visual map of space.[11] The ability of the snake to orient its head toward prey, and the accuracy of its strike, is thought to depend on that multisensory map.

An owl's tectum has a visual map aligned with an auditory map.[12] When the bird hunts, it can aim its strike at the correct location either by the sight of its prey or, when hunting at night, by the sound of the animal rustling in the grass.

Stimulate a monkey's superior colliculus, and a swift, coordinated head-and-eye movement unfolds.[13] The monkey orients toward the mapped location in space. I don't know of any studies applying electrical stimulation to the human superior colliculus, but we are a species of primate and presumably have the same mechanism as in a monkey. When you turn to look at something, especially when an unexpected event causes you to orient in a fast, reflexive manner, your tectum is probably triggering that seemingly effortless, well-coordinated behavior.

All vertebrates use the tectum in more or less the same way, though with some extra bells and whistles, depending on the species. The brain area collects sensory information, picks out the most vivid event happening nearby, and orients the animal, physically pointing the sense organs toward that event.

That kind of orienting is sometimes called "overt attention."[14] It's a simple way to solve a fundamental problem: too much goes on in the environment for the brain to process it all. An animal needs to pick an item of interest and filter out the rest. If you can point your eyes and ears toward one object, then you will automatically filter out other, more peripheral events. The tectum does that job for you. It's evolution's first central controller of attention in the vertebrate brain.

When most people think about the word *attention*, they often mean overt attention. In that colloquial sense of the word, what you are looking at is what you are attending to. If you look away from an object, or turn your back on it, you're not attending to it.

But looking is obviously only one part of the story of attention. A student could be doodling, looking down at a piece of paper on her desk, while still paying *covert* attention to the teacher. Or suppose you overhear two people talking about you. You don't want to turn around and look, but your attention, your processing resources, are covertly centered on that conversation. Or you might be sitting in a chair, daydreaming, your attention focused on something that is not even in the physical world, while your eyes are vacantly looking at the ceiling. In all of these examples, your attention is not where your eyes are looking. This more complex, covert style of attention is outside the job description of the tectum, which only handles overt orienting. With its tectum as its main attentional center, a frog can probably only pay overt attention, not covert attention. It can physically orient itself to objects in the world around it.

THERE IS NO point having attention, any kind of attention, whether overt or covert, if you can't control it. But control is not an easy engineering problem. You need to closely monitor the thing you are controlling. For the first time in this evolutionary story, we will encounter not just cells that can process information, and not just animals that can direct attention, but brain systems that construct an *attention schema*—a bundle of information, called an internal model, that monitors attention. Our evolutionary story is drawing close to something that resembles consciousness—close, but not quite there yet.

A self-driving car needs an internal model of the car. It's not enough for the car's computer to receive information about the outside world and then send signals to the steering wheel and the pedals. The system needs a set of information about the car itself, including its size and shape, the way it handles on the road, and its constantly changing state—speed, acceleration, position. Without a rich, continuously updated internal model that encompasses a good range of information, the car could still have a controller and it could still send out driving commands, but it would probably crash.

This principle of an internal model was first described in the field of engineering.[15] It doesn't matter what is being controlled—it could be something physical, like a car or a robot arm, or it could be something amorphous like the airflow in all the rooms of a large building. To work properly, the control system needs an internal model of whatever is to be controlled. It needs a way to monitor the car, or the robot, or the air currents. In some ways, the internal model is like a general's map on a table with little plastic tanks and soldiers on it. It's a coherent set of information that, usually in a simplified or schematic way, depicts and tracks the thing to be controlled.

The same principle applies biologically. The brain controls the body with the help of an internal model, the so-called body schema, a set of information about the structure and constantly changing state of the body.[16] Sometimes people suffer stroke damage to areas of the brain that build the body schema.[17] If a patient no longer knows about the shape or structure of his arm, he won't be able to move it properly. Simple skills like pointing, reaching, or holding a cup will suffer. But you don't need to look as far as a stroke ward to see the impact of an internal model. If you hang a heavy grocery bag on your wrist by the handle and then try to reach out with that same hand to grasp a doorknob, at first your movements are awkward. Your brain's internal model of the arm is suddenly wrong, because the physical dynamics of the limb have changed. Very rapidly, within a few tries, your internal model begins to learn the new rules and your movements become smoother and more accurate.[18]

From an engineering perspective, an internal model should monitor the present *and* predict the future. If you want to control something, like a shopping cart that you're wheeling down an aisle, you need a way to predict what it's likely to do in the next moment. You create a kind of intuitive simulation of the cart, run the simulation internally, and find out how it behaves. How you steer the real cart—what forces you apply to the handle—will depend on the predictions made by that internal model. Children are terrible at this task, crashing into the supermarket shelves, partly because they haven't learned a good internal model of the cart. They can't predict how their forces on the handle will affect the wobbling of the wheels. Adults, with sufficient practice, tune up an unconscious, intuitive model.

But what about attention? It's arguably the most fundamental process in the brain, and it certainly needs to be controlled. To respond to the world effectively, the brain must focus its resources in a strategic manner on this or that object. And yet, attention can be as wobbly and finicky as a shopping cart, veering off in unexpected directions. By the most basic principles of control engineering, we know that the tectum must use some type of internal model to keep track of attention. My colleagues and I have called that hypothesized internal model an "attention schema," in parallel to the body schema that helps monitor the body. An attention schema is a bundle of information that describes attention—not the object being attended, but attention itself. It monitors the state of attention, keeps track of how it can change dynamically from state to state, and predicts how it may change in the next few moments. A version of an attention schema— information that specifically monitors overt attention—has been found in the tecta of monkeys and cats.[19] On basic principles, the same kind of information is almost certainly present in frogs, fish, or any other animal with a tectum, even if it has not always been explicitly studied.

Now let's return to the humble frog. We know it has a central processor, the tectum. We know it has overt attention, an ability to point its sensors toward a limited part of a larger world. We know it must have an attention schema, because there's no point having attention if you can't control it, and you can't control it without an internal model. The attention schema amounts to a sophisticated self model. The frog doesn't just deploy attention to specific objects in its world. It also, in a way, knows that it is doing so. It has information *about* its own attention.

What exactly does the frog brain know about itself by virtue of having an attention schema?

Let's return to the same thought experiment I used in Chapter 2. Suppose we take out our futuristic, information-to-speech translator, the Speechinator 5000, and plug it into the frog's tectum. Drawing on the information in the attention schema, the Speechinator might say, "Here are some eyes. Here's a body. They move this way and that way, pointing in many directions. Right now they're oriented over here, at that zigzagging black dot. Because they're moving right now, they will soon be pointed in that other direction." The reason for such a literal kind of information is that a frog has a limited kind of attention. Yes, the frog has an attention schema, but it is a schema that describes overt attention only. To the frog, attention means pointing its head and eyes. The internal model that it needs, therefore, is a model of the head and eyes, how they move, and how they align with objects.

Suppose we ask the frog's tectum, through the Speechinator, "But do you have a subjective *experience* of that fly?"

The tectum can only report the information contained within it. It says, "There's a zigzagging black dot. Here are some eyes. Here's a body. They move. They're pointed over there."

In some frustration, we say, "Yes, we got that. But what about *consciousness*? What about a *mental possession* of the fly?"

The frog's tectum replies, "Here are some eyes. Here's a body. They're pointed over there."

The frog's tectum simply doesn't contain the information to answer the questions we're asking. The concept of consciousness is irrelevant to it. Despite having a sophisticated brain, a type of attention, and an attention schema, a frog has no need to contain internal models that describe the self as a conscious agent.

I'm still fond of Elvis and Priscilla. I know they have wonderfully rich behavior, including their croaky mating song. If I spent enough time in the company of frogs, I'm sure I would develop a rapport and an intuitive feel, so characteristic of us social humans, that consciousness must be lurking in the little animals. These are the human, social reasons why people might feel that a frog is conscious. But a frog almost certainly lacks the machinery to model what consciousness is or to attribute that property to itself. It may have *objective* awareness of itself and its surroundings, in the sense that it processes information about its body and its world, but if we could translate its internal information into speech, it would have no reason to lay claim to a *subjective* awareness.

And yet, all the right pieces are almost in place. In my evolutionary account so far, as much as half a billion years ago the earliest jawless fish developed some form of overt attention, a tectum to control it, and probably an attention schema to facilitate that control. Amphibians, reptiles, birds, and mammals all inherited the same system. We all have the same tectal mechanism buried in us. But to find something that we would recognize as consciousness, we will need to take one further step. We must look from overt attention to the much more complex and subtle skill of covert attention, at which birds and mammals are the experts.

CHAPTER 4

The Cerebral Cortex and Consciousness

THE CEREBRAL CORTEX, that wrinkly outer rind so characteristic of the brains of mammals, probably first began to evolve in reptiles more than 300 MYA, in the Carboniferous period[1]—a time of great swampy jungles, when the world was still joining together into the supercontinent Pangaea.[2] The oxygen content of the air was higher than today, and as a result, insects grew to enormous size despite their relatively inefficient respiration systems.[3] Eight-foot-long millipedes and 2-foot-long dragonflies have been found in the fossil record. The period is called the Carboniferous because when the lush jungles eventually died, the vast quantities of biomass turned into our modern coal beds.

The early part of the Carboniferous world would have been teeming with invertebrates and amphibians, the only animals that had yet adapted to life on land.[4] Reptiles evolved from amphibians toward the end of the Carboniferous, as the climate began to dry.[5] They developed a watertight, scaly skin and hard eggshells that allowed them to nest away from water.

They also evolved a part of the brain that is sometimes called the wulst, an expansion of the front-most region, the forebrain.[6] The reptilian wulst is mainly a sensory structure. It contains an organized map of visual space and a map of the touch receptors on the skin, in some ways like the sensory maps in the tectum that I wrote about in the previous chapter. The wulst is like a tectum 2.0, a radical evolutionary rethink, like the iPhone compared with an old-fashioned landline.

Reptiles have an undeserved reputation for stupidity. The "reptilian brain" is popularly supposed to be the ancient core lurking beneath the intelligent parts. But the truth is, reptiles already have the beginning of what became the human cerebral cortex. They have a genuine spark of intelligence. Many reptile species have flexible problem-solving skills and complex social interactions.[7]

Shortly after reptiles appeared in the late Carboniferous, they split into two major groups. The synapsids, the so-called mammal-like reptiles, were at first only subtly different from the sauropsids, the other major group.[8] The synapsids had a slightly different skull structure that gave their jaw muscles a better anchor. Simply put, they were better at eating, which gave them an energy advantage. The best-known early synapsid is *Dimetrodon*, from the Permian period. It's that prehistoric creature in every kid's dinosaur pack, the low-slung one like a croc with a big spiny sail on its back. But it's not a dinosaur. It's a mammal-like reptile, more closely related to us than to *Tyrannosaurus rex*—and it probably stood higher off the ground and less like a belly-dragging lizard than you might suppose from the ungainly-looking plastic toys.

As the mammal-like reptiles gradually transitioned into modern mammals over hundreds of millions of years, the wulst expanded into a layered sheet of neurons that covered much of the rest of the brain.[9] In many mammals, the cortex is a smooth sheet wrapped around the outside of the brain. But in some species, it has expanded so much in surface area and become so packed in that it has wrinkled up like a cloth stuffed into the skull. The human cortex, flattened out, would be fairly close in size and thickness to a large terrycloth hand towel.

Other structures in the brain that have a close connection to the cortex also expanded—especially a large, avocado-shaped object at the base of the brain called the thalamus.[10] It's aptly named. *Thalamus* means "bed" or "foundation" in Latin (originally "bed chamber" in Greek). Every part of the cortex is connected to the thalamus, and most of the information that reaches the cortex passes first through the thalamus, giving it the nickname "the gateway to the cortex." For accuracy, I really should use the awkward techno-jargon "thalamo-cortical system," but for simplicity, I'll stick to the shorthand "cortex" when referring to this specialized circuitry.

As THE MAMMAL-LIKE reptiles evolved their sophisticated cortex over a span of 300 million years, the non-mammal-like reptiles, the sauropsids, found a different path to expanded intelligence. That path led through animals called archosaurs, then dinosaurs, and finally modern birds. The archosaurs were large, low-slung, predatory reptiles with slightly expanded brains compared with their immediate predecessors.[11] Crocodiles, modern archosaurs, are among the most intelligent and behaviorally complex of all reptiles.[12] One can see their intelligence in their crafty stalking of prey, sharing of food, and rearing of young.

By about 230 MYA, in the Triassic period, the archosaurs had produced a specialized subgroup, a bizarre creature that had evolved to run on elongated back legs.[13] For those of you who have an image in mind of four-footed giant dinosaurs, the fossil record shows that all dinosaurs evolved from an original, ancestral type that walked on two legs and probably occasionally touched its hands to the ground. Through later evolution, some species of plant eaters went back down on all fours, while the predatory dinosaurs remained upright on their hind legs.

Everyone knows the cliché that dinosaurs have a brain the size of a walnut. That claim is really libelous. Endocasts from fossil skulls show something of the size and structure of their brains.[14] The largest dinosaurs had correspondingly large brains in total mass. The brain of *T. rex*, for example, probably rivaled the brain of a human in size, although it must have contained far fewer neurons and connections.[15] Dinosaurs would have had the same basic brain organization as crocodiles, including that proto cortex, the wulst. They were not stupid. The theropods, in particular—the two-legged, predatory dinosaurs—may have been the brainiest animals on the planet at the time. With forward-facing eyes that could see in stereoscopic depth, they probably had expanded on the visual part of the wulst to better parse the flood of incoming information.

Birds, as every schoolkid now knows, evolved from dinosaurs. But this popular scientific fact is a simplification of a subtler and stranger truth. Bird flight seems to have emerged gradually throughout the Jurassic period, between about 200 and 145 MYA.[16] It is probably not meaningful to point to a spot in the time line and say, there, that's when birds happened. The blurring between dinosaurs and birds is especially obvious in the evidence from a trove of extremely high-quality, feathered dinosaur fossils, preserved in fine volcanic ash, in China's Liaoning fossil beds.[17] The picture now emerging is that birds *are* theropod dinosaurs. If you had lived in the Mesozoic era, it would never have occurred to you to put birds in a separate category. The world was full of feathered dinosaurs, some large, some small, some toothed and some beaked, strutting around on two legs with their heads bobbing or flying overhead. If you had seen a modern chicken run by or a sparrow fly past, you might have thought, "Look at that! It's yet another subtle variant on the theropod dinosaur!" But you would probably never have said, "Behold, dinosaurs hath given rise to a new class of animal!" In analogy, today nobody would say that bats evolved *from* mammals. Bats *are* mammals that have adapted to flight. Birds, then, are the flying dinosaurs that survived to the present time. And neuroscientists have studied them intensively.

Birds have a wulst that is greatly expanded compared with reptiles.[18] Over the hundreds of millions of years that the mammalian branch was tinkering with the brain, expanding the reptilian wulst into an elaborated cortex, the dinosaur/bird branch was tinkering as well, finding its own ways to modify and expand the same original brain structure.

Although "bird brain" is still considered an insult, birds are intelligent animals. Some species have complex social lives, some are clever hunters, some have excellent memory for food caches. Crows are particularly known for their cognitive intelligence. They can bend wires into hooks and use them as tools to capture food.[19] The famous Aesop fable, in which the crow drops stones into the pitcher, raising the water level until the floating piece of food comes within its reach, is not a myth. Crows can solve that task.[20]

The remarkable intelligence of birds probably depends on their expanded forebrain. Whether extinct dinosaurs were ever as intelligent as modern crows is doubtful—crows represent a recent evolutionary advance—but the same basic brain structures run through the entire reptile/dinosaur/bird lineage.

I will focus mainly on the mammalian cortex, because so much more is known about it and because the only animals today that can verbally attest to being conscious—humans—are mammals. But it's a good idea to keep in mind that other kinds of animals, such as birds or crocodiles, rely on a similar brain system for their rich, complex behavior.

NOW THAT I'VE given a thumbnail sketch of how the cerebral cortex evolved, I'll explain why I think this structure sustains consciousness. To do so, I will need to introduce some details about how the cortex processes information. It's not just a bigger processor stacked on top of the smaller, earlier models like the tectum from the frog brain. It's a fundamentally different *kind* of processor. It's a machine honed to sift a vast amount of information and winnow it down to a small subset. That subset is processed in a deep, thorough manner and ultimately guides behavior. Because of that constant winnowing process, the cortex is, fundamentally, an attention machine.

The cerebral cortex is like the National Football League. In the NFL, teams pass through a series of competitions until the Super Bowl determines the final winner. In the cortex, information percolates through hierarchical levels, continuously subjected to intense, winner-take-all competitions.[21] Regardless of the type of information—visual, auditory, emotional, intellectual—the architecture of the cortex creates elimination rounds of competition. The increasingly selected information becomes ever more deeply processed and ever more likely to have an impact on behavior. Finally, one integrated packet of information wins the cortical Super Bowl and becomes the item that has, at least for the moment, captured the focus of processing.

This elimination-round style of processing has been studied most intensively in the case of vision. Figure 4.1 shows a partial outline of the visual system, based on decades of research in primates, both monkeys

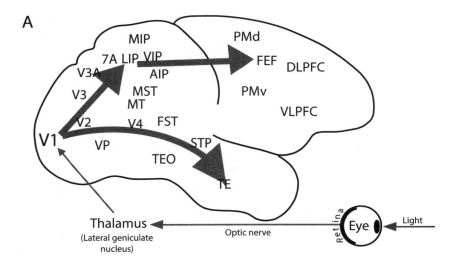

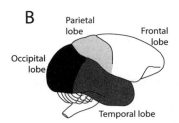

Figure 4.1 Some of the cortical visual areas in the primate brain. (A) A small subset of the major areas involved in vision and the general flow of information through those areas. (B) The four lobes of the cerebral cortex.

and humans.[22] Light enters the eye and forms a patterned image on the retina—the light-sensitive sheet of tissue at the back of the eye. Neurons in the retina become active and immediately begin a local competition. Signals compete against neighboring signals, the stronger ones suppressing the weaker. Signals that correspond to sharp visual contrasts are especially favored.

The information then leaves the eye along the optic nerve, a cable of about 1.5 million fibers, and reaches the thalamus at the base of the brain. The incoming visual information targets one specific part of the thalamus, a bump on its side called the lateral geniculate nucleus (Latin for, literally, "the nucleus sticking out the side that looks like somebody's knee"). Once again, the information is passed through a competitive sieve in the thalamus.

Having passed through the lateral geniculate nucleus, the information then travels along fibers to the first level of the visual cortex, an area in the occipital lobe of the brain, called the primary visual cortex, or V1, where the same type of competition between neurons occurs yet again.

From V1, information flows through a rat's nest of higher-order visual areas—V2, V3, V4, MT, MST, TEO, TE, and so on—an alphabet soup of names. Over the past 50 years, neuroscientists have mapped out the visual cortex in ever-increasing detail, distinguishing dozens of areas and subareas that tile about 60 percent of the cortex, mainly in the occipital, temporal, and parietal lobes. Inside each of these postage-stamp-sized regions of the cortex, bits of information are in constant competition as neurons inhibit their neighbors.

The visual areas have a general organization, trending toward more complexity as information flows from lower-order areas near the back of the brain to higher-order areas closer to the front. For example, neurons in V1, at the start of the hierarchy, pick apart the visual world into short line segments and blobs of color. They perform a simple, superficial, but detailed breakdown of the visual world. In contrast, neurons in TE, an area at a much higher level of the hierarchy, are sensitive to complex images such as faces and hands. They process information about the identity of objects, rather than the details. And yet the information flow is not exclusively up the hierarchical ladder. Signals can flow backward and laterally, passing in any and all directions through the network. Adding to the complexity, the cortex is in constant back-and-forth communication with deeper structures in the brain, especially the thalamus.

As information flows through that processing system, the competition among signals is more like a crooked version of the NFL. The strongest incoming signal does not always win. Other influences can put their thumbs on the scale. When you look out at the world, one item may be brightest or fastest, generating the biggest incoming signal, but it does not necessarily dominate the brain's processing. A signal from the frontal lobe might feed back into the visual cortex and shift the focus of processing to a different, less obvious item, causing it to win the competition instead.[23] When a salient item from the outside wins the competition, it is called bottom-up attention, and when an internal signal puts its thumb on the scale and tilts the competition, it is called top-down attention. But they are both really different aspects of the same underlying process.

The neuroscientist Robert Desimone helped to describe this competitive scrimmage in the cortex and aptly named it "biased competition."[24] I think of it as one of the primary organizational truths of the cortex. The local inhibition between neurons, which creates competition, dominates the machinery of the cortex. It's not an accident that epilepsy is a disease of the cortex. An epileptic seizure occurs when that local inhibition fails.[25] Signals that normally keep each other in check suddenly proliferate and turn into a wild surge of activity that spreads indiscriminately through the cortex. The disease shows just how much inhibition is the essence of the cortex and how drastically the system fails when the inhibition is not sufficient.

In Chapter 2, I described a simple trick in the crab's eye called lateral inhibition, in which nearby neurons suppress each other.[26] The outcome of lateral inhibition is a sharpening of the image. Bright patches register as brighter; dim patches register as darker. Biased competition is the cortex's version of lateral inhibition, blown up a millionfold, expanded from a local competition inside the eye to a teeming world of elimination rounds arranged in a vast hierarchy.

That hierarchy does not stop with vision. In percolating through the interconnected net of brain areas, visual information ultimately comes up against other forms of information. Touch, hearing, movement control, abstract thinking—vast domains of information converge in a set of areas that may be the highest-order, most integrative networks in the brain. These areas tend to cluster mainly in two lobes, the parietal lobe and the frontal lobe, and are consequently called, rather unimaginatively, the parietal-frontal networks.[27] Information that reaches these networks has won the Super Bowl. It has achieved what the philosopher Daniel Dennett has called "fame in the brain."[28] The cortical machine of elimination rounds has winnowed down the world to a few items, which are now influencing the most central networks in the cortex and stand a high chance of driving behavior and entering memory. But the focus of one moment may not be the focus of the next. The competition is always shifting, always dynamic, as new bits of information win through the neuronal networks to the highest levels of processing. The cortical system is restless, by the very nature of how it operates. The running competition is intrinsically unstable.

That restless, cortical winnowing process is our most powerful type of attention. It's the most sophisticated tool the brain has for understanding the world. Sometimes people ask me why I built a theory of consciousness centered around such a specific, seemingly narrow topic as attention. But attention is not a narrow topic. It is the essence of how the cerebral cortex apprehends the world.

THE TECTUM, that primitive, half-billion-year-old brain structure that I described in the last chapter, dominates overt attention. It can point your eyes and ears, like satellite dishes, to collect more information from one part of the world. The evolution of the cortex, however, opened up a new approach to processing the world: covert attention. You can attend to something that is not directly at the center of your gaze.

I do not mean that the cortex has nothing to do with eye movement. It plays a large role in helping to coordinate the eyes and head.[29] But cortical attention is not bound to where the eyes happen to be pointing. You could be looking nervously down at your shoelaces while your cortical attention is nailed on your boss standing in front of you. You could be paying attention to an itch in the small of your back, your whole world reduced to that unpleasant sensation, while at the same time you're smiling and looking directly at a friend, hearing nothing she says. You could be staring at the printed page of a book while your attention is focused on a memory from yesterday, an internal thought that has nothing to do with the direction you're looking.

The difference between overt attention and covert attention is simply this: overt attention is grasping an object with your sense organs; covert attention is grasping an object with the massive computational machinery of the cortex. You can "point" that machinery toward any item, whether it is a concrete object in front of you, something off to the side, or something more internal like a thought or an emotion.

Scientists have sometimes compared covert attention to a spotlight.[30] In that analogy, the spotlight has a bright center focused on one or two objects and an extensive, fuzzy fringe that might take in other objects. Of the hundreds of objects around you at each moment, your cortical spotlight illuminates a subset, while the rest, the objects that are entirely unattended, effectively do not even register with you. The analogy is a little strained, however, when you consider that a spotlight can only move across literal space, whereas covert attention can move through dimensions that have nothing to do with space.[31] You can look at a Mondrian painting and focus your attention so much on the colors that you fail to pay attention to the shapes and can't even remember them later, even though the colors and shapes were both at the same location in space. That kind of attention isn't much like a physical spotlight. Given those considerations, the spotlight analogy has fallen out of favor. I personally like the analogy, but only if the spotlight is understood to rove about a space of abstract dimensions and not just the three dimensions of physical space.

The cerebral cortex creates that inner spotlight. It allows us to explore a nearly infinite, multidimensional landscape over which our focus of processing ranges, from the most concrete and immediate objects to the most abstract ideas.

You can't have deep processing without a shifting focus of attention. If you could deeply process all available information simultaneously, you would need a brain the size of a planet. Evolution found an efficient way to use a limited amount of brain power and still intelligently process the world. The solution was to subject information to such a fierce competition that only a small amount is fully processed at any one time and to build into the system a sophisticated controller for shifting and adjusting that focus of processing. Attention is the key that unlocked a complex understanding of the world.

SCHOLARS WHO WRITE about the evolution of consciousness tend to emphasize a gradual increase in the complexity of the brain. It is intuitively tempting to think that complexity makes consciousness.[32] In that view, somewhere in the process of evolution, the nervous system became so complex that it crossed a threshold, woke up, and gained subjectivity. If that is true, then the question of consciousness turns into a matter of finding the threshold—always a slippery-slope proposition.

We might start with the assumption that fish are dumb automatons that can't possibly have consciousness. Then an expert on fish sets us straight, describing unexpectedly rich behavior,[33] and we're encouraged to conclude that fish might just be conscious after all. Bees are mere insects with tiny brains; and yet they, too, have such computational complexity that, in some ways, they rival what our pet dogs can do.[34] I know someone with a pet tarantula who insists that the animal has so much complexity in its behavior, so much personality and temperament, as to suggest consciousness. Then a whole set of human emotional biases begins to creep into the discussion unannounced. If bugs are conscious, then maybe nobody should be smashing them, so perhaps I'll draw the line conveniently a little higher up the scale of nervous complexity and assume that bugs aren't conscious after all—which leaves me with some contradictions that sneak in through sheer human irrationality.

The difficulty with the complexity argument is that the more you look into it, the more slippery and arbitrary that slope becomes and the more subject to human whim. Single-celled animals have their own complex methods of processing information through interacting chemical signals. If you're an expert on amoebas and spend your life spying on them through a microscope, learning the subtleties of their ways, you might champion amoeba consciousness. Plants have their own electrochemical messaging systems, processing information about the outside world, so maybe plants are conscious. The genome in the nucleus of a cell, together with all the massive chemical machinery that handles the genetic code, is quite an information-dense computer. Maybe the cell's nucleus has its own consciousness.

But why stop there? Rocks and water and single electrons are teeming with information and fluctuating states, sensitive to their outside environments. Why aren't they conscious, too?

Once you start with the intuition that consciousness arises naturally from complex information processing, it's hard not to slip into panpsychism, the belief that everything in the universe is conscious to at least some degree.[35] By replacing consciousness with information and complexity, properties that are literally everywhere and in everything, we are left to slide up and down the slippery scale.

The reason I suggest that the cerebral cortex sustains consciousness is not because it's the most complicated machine in the known universe—although it is that. The reason is much more specific.

Let's say you're looking at an ordinary object—an apple. It's sitting on a table in front of you, and you have a clear view of it. The visual information about the apple flows through your processing system, from the retina through the cortex. Suppose the signals related to the apple win the elimination rounds of competition. As we look higher up your cortical hierarchy, we find that the apple information is processed more deeply, while competing signals are suppressed. The apple information reaches the central networks in the brain, the parietal-frontal networks. Ten thousand other possible chunks of information have lost the competition of the moment. Sounds and sights around you, the feel of your clothes on your body, memories, ideas, emotions, all of these have lost to the apple. At this one moment in time, the visual image of the apple, and perhaps a few other items, dominate your cortical attention machine, whereas other, rival information is processed to a much lesser extent, in much more limited regions of your cortex.

In one major theory of consciousness, called the global workspace theory, we're done.[36] We have a complete account of your conscious experience of the apple. I'll discuss that theory again in a later chapter, when I compare the attention schema theory with some of the main alternatives and suggest that many of the theories are not actually rivals, but instead are deeply related to each other. In the present chapter, I want to briefly point to the global workspace theory because it provides a useful perspective on attention in the cerebral cortex.

In the global workspace theory, information percolates through the cortical system. It is selected and boosted by attention until it crosses a threshold and influences widespread networks throughout the cortex. It has won that highest level, the Super Bowl of attentional competition in the brain. By entering that state of widespread influence, it has reached the global workspace. In that theory, information in the global workspace *is* information in consciousness. Why the global workspace has the property of consciousness associated with it is not explained. The theory is, in a sense, descriptive but not yet explanatory.

To show how the global workspace theory is incomplete, let's ask your brain, as it is outlined here, about this moment in its world. In previous chapters, for nonverbal animals, I relied on the Speechinator 5000, a hypothetical machine that can be plugged into a brain and will translate the information it finds into speech. Here we don't need the hypothetical device, but I want to make sure we follow the same logic. The speech machinery in the human brain can verbalize information contained in the global workspace. The fact that speech depends on internal information may seem obvious, but it is easy to get wrong. Our casual intuition tells us that when people speak, they are merely expressing their inner experiences. But in reality, speech is a type of information output. If a particular set of information is missing from inside the system, then it can't be expressed verbally.

So if I ask you, "What's up? What's in front of you?"

You can reply, "An apple."

If I press you on the details, you can provide them—color, texture, shape, location. All of that information is present in your global workspace.

But suppose I ask you, "Are you conscious of the apple? Do you have a subjective experience?" Now we've hit a snag. Thus far, in discussing cortical attention, I have not explained how or why your global workspace, or any part of the cortex, would contain information about consciousness. It contains information about the apple and therefore can report visual details; but on what basis could you possibly provide a response when asked about conscious experience? The very notion of consciousness would mean nothing to you.

We're still missing a crucial part of the explanation for consciousness. As I explain next, we need to add an attention schema.

I'VE DESCRIBED HOW the cortex is, at its heart, an attention machine. But there is no use in having attention if it flaps around randomly or is driven entirely by externals like the brightness or loudness of stimuli. It needs an internal control system, and a controller cannot operate without an internal model. The attention schema is that internal model, a set of information about the process of attention itself.

To get across the importance of an attention schema, I'll use an analogy that may seem a little far-fetched at first, but will prove apt. I'm thinking of a memorable scene in the movie *Butch Cassidy and the Sundance Kid*. The Kid, played by Robert Redford, is auditioning to be a deputy. He's asked to shoot a target and told to stand upright and stiff with what passes as good shooting form. Somehow, he can't hit the target. His aim is terrible. Finally, in frustration, he says, "Can I move?" Then, with his body wiggling and the gun flapping, he shoots two bullets and nails his target each time. "I'm better when I move," he mutters.

That scene is a beautiful illustration of control engineering. You should keep most of your degrees of freedom loose and flappy if you can. You don't want to try to monitor and control everything. You don't need a model of every last detail. Instead, a good controller monitors, models, and constrains the crucial outcome—in the Sundance Kid's case, whether the bullet hits the target.

Another classic example of control engineering is hammering a nail. With a little practice, people are good at swinging a hammer and consistently hitting the head of the nail. But how exactly do we control the hammer? One approach might be to monitor and control each separate degree of freedom. Your shoulder rotates, your elbow straightens, your wrist flicks. If you can optimize each one of those movements and hone it to reliable perfection, maybe you'd make an expert hammerer. But as it turns out, that is not what people do. Instead, we focus on the head of the hammer as it strikes the nail.[37] We monitor and control that one degree of freedom, because it is the only outcome that matters for the task. If you film people hammering and track their movements in detail, you find that the shoulder rotates in variable ways, the elbow wobbles unreliably, the wrist movement is a little different every time, and yet the head of the hammer manages to hit the nail square on (at least once you've become an expert at the task). The control system has no need to model or directly constrain the details as long as it can control the crucial output—hitting the nail on target, in this case. The details are free to vary as long as they collectively add up to the right output.

The Sundance Kid's target practice and the subtle art of nail hammering are both straightforward examples of how the brain uses internal models. The brain needs to model the essence, not the incidental details. Otherwise, like the Sundance Kid forced to monitor and constrain too many intermediate details, the control system would break down and fail to accomplish the task. When the brain controls attention, directing it here, holding it there, shifting it quickly or slowly, broadening it or sharpening it, what kind of internal model does it use? What aspects of attention are simulated by that model?

In the case of overt attention, which I described in the previous chapter, the brain can keep track of attention by monitoring a physical body part—the eyeball, for example, as it moves and points to different objects. But in the case of covert attention, no physical body part is involved. When covert attention moves from one item to another—from the apple, to a sound, to a memory—nothing tangible is actually moving through space. Instead, billions of neurons are changing their activity state in subtle ways. At a microscopic level, covert attention is still a physical process, but a useful attention schema would not model the details of neurons, inhibitory connections, elimination rounds of competition, cortical hierarchies, or networks spanning the parietal and frontal lobes. The brain has no need to know all the internal, mechanistic details that I've outlined so far in this chapter. Instead, to be useful, the attention schema must depict something simplified that cuts to the pragmatic essence of covert attention.

Now, finally, we reach the central proposal of the theory. My colleagues and I propose that the cortical attention schema has a particular form. The information within it provides a cartoonish account of how the highest levels of cortical attention take possession of items. There is no simple, physical, roving eyeball, as in the case of overt attention. Instead, that cartoonish account describes an essence that has no specific physical substance but that has a location vaguely inside you, that can take temporary possession of items—of apples and sounds and thoughts and memories—and that restlessly moves about, searching, seizing some items and dropping others. When that ethereal mental essence takes hold of an item, it has the property of making the item clear to you, real to you, vividly present—in other words, it turns the item into an experience. It also has powerful consequences. It enables you to understand the item, to respond to it, to talk about it, or to remember it so that you can choose to act on it later. It *empowers* you to react.

That amorphous power inside you is a fictionalized, detail-poor account of cortical attention. Not low-level attention, like the competition between visual signals in V1, but the highest level of attention, in which an item like an apple can win the cortical Super Bowl and have an impact on your behavior.

When I ask, "What is your mental relationship to that apple?" your verbal machinery can tap into the information available in your cortical networks, information that has reached your global workspace. It not only draws on visual information about the apple, but also on information in the attention schema about an amorphous power inside you. Those two sets of information are linked together into an integrated whole, something like a complete dossier relevant to the apple at that moment. Based on that pool of information, you can say, "As I look at that apple, I have a mental possession, a *conscious experience*, of its redness."

Suppose I ask, "But what do you mean by a conscious experience? What are its specific, physical properties?" You can't easily answer that question. The attention schema, being detail-poor, lacks a description of any specific physical properties of attention.

I ask, "Can you scratch your conscious experience and measure its hardness? Can you put it on a scale and measure its weight? Can you heat it and measure its combustion temperature? What physical measurements can be made of it?" You might answer, "It doesn't have any of those physical characteristics. In that sense it's nonphysical, or metaphysical. It's simply a mental experience—my mind's way of grasping something. Don't you know what consciousness is?"

Logically, the brain cannot put out a claim unless it contains the information on which the claim is based. The attention schema theory focuses on the information set on which the claim of subjective experience is based. Because your brain constructs a schematic model of cortical attention, you know what consciousness is and you think that you have it. You can answer questions about it when asked, and when reading a book about consciousness, like this one, you know more or less what property I'm referring to. Without an attention schema, you would be missing the information required to do any of those things, and consciousness would be meaningless and irrelevant to you.

In my view, the attention schema theory of consciousness has some inevitability to its logic. First, we know that the cortex uses covert attention. Second, we know that it needs to control that attention. Third, we know that the brain must have an internal model of attention in order to control that attention. Fourth, we know that a detailed, fully accurate internal model is at best wasteful and at worst harmful to the process, and so, this internal model of attention would necessarily leave out the mechanistic details. Therefore, and fifth, an attention schema would depict the self as containing an amorphous, nonphysical, internal power, an ability to know, to experience, and to respond, a roving mental focus—the essence of covert attention without the underpinning details. From first principles, if you had to build a well-functioning brain that had a powerful, cortical style of covert attention, you would build a machine that, drawing on the information constructed within it, would assert that it has a nonphysical consciousness.

That cortical machine, of course, would not know that its subjective conscious experience is a construct or a simplification. It would take the nonphysical nature of conscious experience as a reality, because—somewhat tautologically—the brain knows only what it knows. It is captive to its own information.

THINK OF ALL the aspects of brain function I've covered in the last three chapters that are not, themselves, enough to explain consciousness. Complexity, by itself, isn't enough. The most complex brain in the world might have no information in it relevant to consciousness. The overt type of attention controlled by a frog's tectum is also not enough. A frog can point its head toward a fly or a threat and not have any information inside its brain relevant to consciousness. Even a more sophisticated, cortical style of attention isn't enough. When information about an apple wins the cortical competition and reaches the global workspace, that state by itself gives you no traction on consciousness. There is still no logical reason why a machine of that type would know anything about consciousness, or think it has consciousness, or make claims about consciousness. You need one extra piece. You need an internal model that describes cortical attention. That attention schema seems like such a small addition, and yet only then does the system have the requisite information to lay claim to a subjective experience. With that piece in place, consciousness finally becomes relevant to the machine.

EVERYONE WANTS TO KNOW—which animals are conscious? In the past few chapters, I described how fragmentary components of consciousness may have existed in a long history dating back half a billion years or more. Figure 4.2 shows how those components may have come together, one by one, during evolution. If the attention schema theory is correct, then consciousness as we humans understand it probably appeared early, maybe as early as 300 million years ago. The underlying brain structures began to emerge in reptiles, are probably present in birds, and are definitely present in mammals. These three groups may vary in the sophistication or richness of their covert attention and their attention schemas, but they may all have some version of what we would recognize as consciousness.

And yet I am not done with the evolutionary story. The most human part has yet to come. We not only build rich, descriptive models of ourselves, but also reflexively attribute consciousness to each other, creating a social ecosystem. We see consciousness in other people, in our pets, in our toys, in the vast invisible spirit world of gods and ghosts that we project into the spaces around us. In the next chapter, I'll take up that crucial social use of consciousness that has expanded so radically in our species.

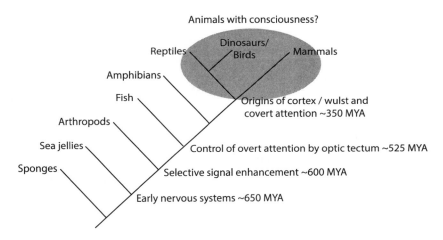

Figure 4.2 Possible evolution of consciousness, from sponges to mammals.

CHAPTER 5

Social Consciousness

WE HUMANS DON'T just guess at each other's mind states. Neither do we tote up deliberate observations and then intellectually figure each other out. Well, we do that sometimes, but not with great success. Instead, we have a well-tuned intuition. We seem to *know* what other people are thinking and feeling. Sometimes the knowledge is so palpable that it seems as if we can directly sense each other's thoughts and emotions like emanations. We can't, of course. But millions of years of evolution have given us an ability to read subtle cues and construct rich models of each other's minds—which we do intuitively rather than explicitly.

We attribute emotions, intentions, agendas, beliefs—the whole range of mental content—to each other. That gorgeously complicated human process of reconstructing someone else's mind is called *theory of mind*.[1] It isn't a theory in the intellectual sense; it's automatic and inevitable. We can't help it. But when we build these models of each other's minds, one component is particularly important: reconstructing someone else's attention.

How can I know whether you plan to reach for that apple unless I know whether you're paying attention to it in the first place? And even if I know you're attending to the apple, how can I make predictions about what you'll do or say next unless I understand the consequences of attention? My first step in reconstructing your mind is to understand that a mind is a thing that has a focus, that the focus can be broad or narrow, depending on circumstances, and can move around from item to item, and that the focus has predictable consequences. Without that, I wouldn't have much of a theory of mind. I need more than a theory of the *contents* of your mind; I need a theory of what a mind *is* in the first place.

If I'm standing in front of you, it's easy for me to see where you're looking. A whole science has sprung up around how the brain processes someone else's direction of gaze.[2] But tracking your overt attention isn't nearly good enough for me to reconstruct your mind state. I need to understand your *covert* attention. I need to draw on every clue I can from the entire context, including your body language, your facial expression, your words, and my general knowledge of you. Regardless of where your eyes happen to be looking and whether or not I can even see your eyes, I need to reconstruct what might have percolated up through your cortical hierarchies and reached your highest levels of processing, and I need to understand how those highest levels of processing might impact your behavior.

I guarantee that nobody ever looked at another person and had the immediate, intuitive thought, "That person's cortical visual pathways are engaged in processing multitudinous stimuli at the present moment, of which the neurons representing the shape of the apple have had their activity increased due to a boosting signal from regions of the frontal lobe, and in consequence those boosted neurons have partially inhibited the neighboring neuronal representations through lateral inhibition dependent on local interneurons that use the neurotransmitter gamma-aminobutyric acid. . . ." I could go on with the technical jargon. But the truth is that nobody ever attributes true, physical, neuronal attention to another person. We don't need to, especially on that level of detail. Instead, my brain constructs a much more schematic, efficient model. I intuitively understand,

"Right now that person's consciousness seems to have taken hold of the apple, with many possible consequences."

We are back again to consciousness as a simplified, useful model of attention. But this time, the attention schema is being used for social intelligence: to model someone else instead of oneself.

SALLY AND ANNE walk to the park with two covered picnic baskets. Sally puts a sandwich away in basket A and leaves to visit the restroom. While she's gone, Anne sneakily switches the sandwich to basket B and closes the covers again. When Sally returns, which basket will she look in first to find her sandwich? This simple question has become the go-to test for a person's theory-of-mind ability.[3]

To pass the test, you need to keep track of the knowledge in Sally's mind. What she knows about the sandwich's location is correct when she first puts it in basket A, but then becomes incorrect when the sandwich is switched to basket B. You can't solve the task without a concept of Sally's mind as a separate thing that contains information, possibly false information, which determines Sally's actions. For that reason, the test is also sometimes called the false belief task. The correct answer is: she'll look in basket A first and discover that her sandwich is missing.

Children before the age of about 5 can't consistently give the right answer.[4] To them, if the sandwich is in basket B, that's where Sally will look. Why would she open a basket that doesn't contain what she's looking for? As children mature past age 5, their social thinking tunes up and the task becomes intuitively easy. By the time we're adults, we're generally relatively good at keeping track of other people's mind states.

Chimpanzees show some evidence of solving the false belief task.[5] Suppose you stage the Sally-Anne scenario in front of a chimp. As the chimp watches, Sally places a piece of fruit in a box, then leaves. Anne switches the fruit to box B. Finally Sally returns to retrieve her fruit. When you track the chimp's eyes, it tends to gaze more at box A, where

Sally thinks the fruit is located and where she's presumably about to reach, rather than at box B, where the fruit is actually located. The chimp must be keeping track of the contents of Sally's mind and anticipating her actions.

Ravens may also be able to solve a similar task.[6] These birds routinely store food for later, but they don't like it when other birds steal their treats. Suppose a bird hides a piece of food while a second bird watches. The watcher then leaves. While the watcher is gone, the hider will craftily rehide the food, perhaps to prevent it from being stolen in the future. It seems as if the hider understands that the watcher has gleaned knowledge about where the food is hidden, and when the watcher returns, it will now have a false belief and will consequently look in the wrong place.

Not many nonhuman animals can pass the false belief task. Even the few exceptions I've noted are controversial. But in my opinion, concluding that other animals must lack a theory of mind is premature. The false belief task is too high a bar. Keeping track of multiple boxes and their switched contents is intellectually complicated, like a difficult shell game. No wonder only humans can solve it consistently. I'm interested in something simpler: the concept of mind. We all understand that Sally has a mind, and a mind is something that can contain information and will guide behavior based on that information. Do other animals have the same intuitive understanding? Do they know what it means for someone else to be conscious of something?

Scientists who study animal behavior prefer simple explanations. Rather than suppose that an animal has any concept of someone else's consciousness, it seems easier to presume that the animal has merely learned a set of simple rules. For example, maybe a zebra doesn't need to know that a lion has become aware of it. Instead, the zebra can just run away from anything big and toothy. Stimulus in, response out. If the zebra contains a large enough library of these associations, it can presumably get by. I would point out, however, that this kind of hypothesis, so typical of stimulus-response psychology, is actually naive. A vast library of learned associations is not the easiest or most efficient method of navigating a complicated environment. From a computational point of view, a model-based approach is probably easier, because one model can serve a broad range of circumstances. It may be a simpler, less computationally expensive solution for the zebra to build a schematic model in which the lion has a mind, the mind can take possession of items in the world, and once having done so, the mind can then guide the lion's actions.

It may seem far-fetched to suppose that a zebra "understands" the consciousness of another animal, but it seems that way only because we humans think of consciousness as a noble characteristic, special to us and rich with cultural associations. Zebras shouldn't have the poetry or complexity for it. But that's just our ego talking. I suggest that consciousness is an ancient part of theory of mind; it is a simple, efficient model that is used to make predictions about the behavior of animals and is likely to have evolved long before humans. I would not be surprised if zebras, other mammals, birds, and maybe even some reptiles use that convenient construct of consciousness, perhaps with varying levels of complexity, to predict each other's behavior.

It's a favorite trope that we humans have a superior consciousness. We tend to think that other animals are not conscious at all or that their consciousness is less developed. This perspective dovetails with the popular idea that consciousness emerges from complexity. Because we have the most complex brains of all animals, we must also have the most vivid consciousness. Yet of all the mental talents that we humans like to brag about—math, language, tool use, and so on—consciousness may be one of the most primitive and least special to us. I accept that the contents of consciousness, the thoughts and ideas and beliefs, the insightful perceptions, the knowledge of mortality, are probably more sophisticated for humans than for other animals. But the fact of consciousness itself, our ability to have a subjective experience of anything at all and to attribute subjective experience to others, is of such basic utility that it may be shared across a vast range of the animal kingdom. If the attention schema theory is correct, then it is certainly not unique to humans.

To GAIN BETTER INSIGHT, I often find it useful to take an engineering perspective. Let's think through how we might engineer a machine that can predict human behavior, moment by moment. The exercise will show just how difficult a task the brain solves, while at the same time revealing some of the underlying principles. It will also point the way toward the later chapters of this book, when I discuss whether we could ever build artificial consciousness.

Suppose a person (we'll call him Kevin) walks into a room. Unseen, a camera eye watches him, and a microphone listens in. The camera and microphone are connected to a computer system, the Predictor 5000, whose job is to give a running prediction of what Kevin will do.

The room contains the following items: a white-powdered doughnut in the middle of a table in the middle of the room, with an overhead light shining brightly on the doughnut; a small puddle of water on the floor in front of the table; and a cell phone on a shelf in the corner of the room, where the light is dim.

The first task we'll give our prediction engine is to identify the affordances in the room. The influential psychologist and naturalist J. J. Gibson coined the term *affordance* in the 1970s.[7] He realized that when an animal or a person looks around its world, the real job of its visual system isn't to take in the world as it is, like taking a picture, but to identify specific opportunities for action. Those opportunities are called affordances. A fly provides an affordance to a frog; the frog can grab it and eat it. A branch provides an affordance to a bird; the bird can perch on it. A doorknob provides an affordance to a person; the person can turn it.

Right away we can see the great complexity of building a working, behavioral prediction engine. The Predictor 5000 needs expertise in human behavior in order to look at a person, then look at a puddle, and retrieve the information that the person might step over the puddle. Imagine that task repeated for every possible object a person could encounter. An effective prediction engine would need a vast amount of miscellaneous background knowledge on human habits. Although that kind of knowledge is not intrinsically complicated to implement in artificial intelligence—it's a matter of recognizing objects and associating them with potential actions—still, there are an awful lot of associations to learn.

To make the problem more difficult, each object can have a large number of affordances. The room may seem spartan with its three objects, but it contains a buffet of hidden possibilities. Kevin might step over the puddle; he might jump into it and make a splash; or he might take out a handkerchief and mop it up. He might toss the table over in a rage; he might neatly move it to another part of the room; he might climb on it; or he might hide under it. He might reach for the doughnut and eat it; he might throw it on the floor and stomp on it; or he might put it to his eye and look through it like a monocle, as a joke. He might pick up the phone and try to make a call; he might pocket the phone surreptitiously; or he might do something as simple as walk toward the phone to look more closely at it. The prediction engine is faced with a huge number of possible affordances.

To simplify the task, we'll give our Predictor 5000 some information on prior probability—general information about how an average person behaves. Most people don't normally jump into a puddle, throw over a table, or stomp on a doughnut. Those are low-probability events. Stepping over the puddle and eating the doughnut are higher-probability actions. With enough data on human behavior, we can figure out a numerical probability for the average person performing those particular actions when faced with a puddle or a doughnut.

But even after we front-load the machine with that background information, we still don't have a good prediction engine. The information, so far, reflects the average person. We don't know how our particular specimen, Kevin, will act on this occasion. Let's help out our prediction machine by giving it the whole apparatus of a traditional theory of mind. For example, if I know Kevin hasn't eaten for 10 hours, I might suppose a higher probability of him snarfing down the doughnut. If I know he has diabetes, I might suppose he won't eat it. If I know Kevin has an impulse control problem or is really angry at the moment, I might suppose he has a higher chance of stomping in the puddle. The philosopher Daniel Dennett calls this type of information about a person the "intentional stance."[8] We are always looking at other people and automatically asking ourselves, "Can I figure out that person's motivations so that I can predict what that person might do next?"

Researchers have now begun to build artificial systems to try to tackle the problem of behavioral prediction using intentional stance.[9] Being able to guess someone else's intentional stance presumably requires storing up an intricate knowledge about human nature—about typical human motivations and especially about emotional facial expressions. But I want to sidestep the issue of the intentional stance and get at something simpler and, I think, more fundamental.

Let's suppose I've taken the available statistical information on general human behavior, combined it with specific information on Kevin and his particular intentional stance this morning, and rolled it into one set of numbers. Each affordance in the room now has a probability assigned to it. Eating the doughnut: 30 percent. Carefully stepping over the puddle: 50 percent. Picking up someone else's phone that has been left on the shelf: 3 percent. I've done all the background work normally considered the substance of social cognition and given that information to the Predictor 5000. It seems I've done its job for it.

Yet with all of that useful information front-loaded into it, the machine still can't predict Kevin's behavior on a moment-by-moment basis. It needs information about a crucial, hidden variable that changes in real time: how Kevin is focusing his attention. Kevin's processing resources are constantly shifting, moving about the room. As a result, the probabilities for acting on the doughnut, or the puddle, or the phone are constantly in flux.

Take the case of the doughnut. Throwing the doughnut on the floor and stomping on it has such a low prior probability that the action is discarded by the prediction engine. Using it as a monocle is likewise discarded. We're left with only one realistic or likely affordance with respect to the doughnut: picking it up and eating it. But the probability of that act fluctuates over time. To try to predict that fluctuation, I'm going to get slightly mathematical, using a framework called Bayesian statistics. If you're not interested in math, you can skip this part, but I promise to make it as straightforward as possible.

Suppose that, given everything we know about people in general and about Kevin in particular, we think that his overall probability of eating that doughnut is about 30 percent. That number is called a prior probability, and it's that prior probability that we have front-loaded into the machine. Let's label it as P_{prior}. Now suppose that our prediction machine computes a second number, which can also vary between 0 and 100 percent. This second number is an estimate of the amount of attention that Kevin is directing toward the doughnut. We'll call that number C_1, where the subscript 1 refers to object 1, the doughnut. Later we'll look at C_2 and C_3, referring to the amount of attention Kevin directs toward the puddle and the cell phone. C_1 is constantly fluctuating in time. Most of the time, Kevin is paying little or no attention to the doughnut, and C_1 is close to 0. Occasionally, his attention to the doughnut may flicker up for a moment, and C_1 will rise a little; or his attention to it may surge, and then C_1 will temporarily peak at around 100 percent.

The more Kevin's attention is focused on the doughnut—hence the larger C_1 becomes—the more likely he is to act toward the doughnut. C_1 is a kind of permissive variable, unlocking the possibility of action. Now comes the only equation I'll throw at you. Suppose you ask the Predictor 5000 machine, "What is the probability at this particular moment in time, let's call it P_{action}, that Kevin will eat that doughnut?"

The machine estimates that probability with a simple computation:

$$P_{action} = C_1 \times P_{prior}$$

That's it. Multiply the prior probability by the estimated amount of atten-tion, and you have a way to predict Kevin's behavior, moment by moment. As long as he's paying no attention to the doughnut, then $C_1 = 0$, and therefore $P_{action} = 0$, and the machine predicts he won't eat it. As soon as his attention to the doughnut spikes, his estimated probability of eating it spikes. Even at the peak of that spike, his probability of eating the dough-nut never goes above P_{prior}, which may after all be quite small, since people don't often pick up and eat random doughnuts. As his attention to the doughnut drops back down again, his probability of eating it also subsides back toward 0. The usefulness of this kind of computation is that it takes the more standard theory-of-mind approach, which tends to operate in the framework of static snapshots in time—if Kevin is handed a dough-nut, will he eat it?—and puts it in a framework that can accommodate continuous, dynamic changes of attention from second to second.

The task for the machine is to estimate that constantly changing num-ber, C_1. But our artificial prediction engine does not have direct access to Kevin's brain. Kevin's attention is the result of a highly complex set of neuronal interactions, hidden away inside his skull. The machine merely has a camera trained on Kevin and a microphone to pick up sound in the room. What it needs are some simple heuristics that can turn its limited observations into an estimate of Kevin's state of attention.

To help our machine, let's harness a few well-established, scientific insights about how attention works. First, the doughnut is white and under a bright light—it has high stimulus salience, a perceptual pop-out effect. A high-salience doughnut means that Kevin is likely to pay more attention to it. Hence, the machine might raise its estimate of C_1. Second, the doughnut is alone on an otherwise empty table, and attention depends inversely on clutter or visual competition. Again, based on that clue, the machine might increase its estimate of C_1. Third, the direction of Kevin's gaze supplies useful information. Gaze is not a perfect indicator—he could be staring straight at the doughnut and yet occupied by something else, such as listening intently to a noise from the corridor behind him or thinking hard about his plans for tomorrow. But on average, gaze is an important consideration when trying to estimate someone's attention. Fourth, his facial expression could give clues. If his face changes suddenly from a neutral expression to a more focused expression just as his eyes point at the doughnut, then the machine would have a strong basis for computing that the value of C_1 has just increased.

Throwing all these different considerations into the mix—the relationship of attention with saliency, clutter, gaze direction, and facial expression—the machine can estimate a time-varying value for C_1, the amount of processing resources Kevin is directing at object 1. That estimation fluctuates over time as the machine takes in changing information. Using the value of C_1, our machine can compute a running probability that Kevin will pick up and eat the doughnut: right now he might; now he won't; definitely won't; nope; yes, now he might again.

The situation gets a lot more interesting when we start to consider the other objects in the room. Take the puddle on the floor. The machine computes a value for C_2, which represents the estimated amount that Kevin is focusing his attention on object 2. The puddle is definitely not in a central, obvious position in the room. It's in the shadow of the table, so it's not bright or glittering. In other words, it's not salient. Suppose Kevin's gaze is never directed toward the floor. He may still have processed the puddle in his peripheral vision, but as a first approximation, the machine may reasonably estimate a low value of C_2—and therefore a low probability that Kevin will step over the puddle as he walks into the room toward the table. If he's not paying attention to it, he'll splash right into it.

The two numbers so far, C_1 and C_2, depend on each other. They are intertwined. Kevin's processing resources are limited, and so as C_1 increases, C_2 must decrease. As he pays more attention to the doughnut, he's less likely to attend to the puddle, and vice versa. If his attention is eagerly fixed on the doughnut as he walks toward the table, he's likely to tread in the puddle. Computing C with respect to each item in the room therefore requires a deeper model that can take into account the competitive interactions between items.

The machine can also compute C_3, an estimate of how much Kevin is attending to the third object, the phone. Initially, the machine computes a low value for C_3 because the phone is tucked away in a shadow at the back of the room, and Kevin's gaze is not directed at it. Now the phone gives a single ring, gaining higher salience. Registering that change in salience through the hidden microphone, our prediction machine can compute a sharp increase in C_3. Kevin's attention probably surged toward the phone at that moment. Even if his gaze is fixed elsewhere, given the intense salience of the stimulus at that moment in time, C_3 should be high. As a direct consequence, C_1 and C_2 must drop at the same instant. The three numbers are in a constant, competitive dance with each other. At the moment the phone rings, Kevin's probability of picking up the doughnut suddenly drops close to 0. Moreover, Kevin's attention, drawn to the phone when it rings, has some stickiness or viscosity. It will tend to remain focused on the phone for at least half a second after the ring and then fall away gradually along a curve that is typical of human attention. Attention does not move instantaneously from object to object. To predict behavior optimally, the computation of C needs to incorporate the approximate viscosity of human attention.

Putting all of this complexity together, our Predictor 5000 must compute the ever-changing values of C_1, C_2, and C_3. The computation relies on basic information about how human attention works. The machine draws on clues such as the salience of different stimuli in the environment, the clutter in the environment, where Kevin's eyes are directed, his facial expression, and the temporal dynamics of human attention. As the machine calculates the time-varying values of C_1, C_2, and C_3, it can estimate the ever-changing probability that Kevin will engage in actions afforded by the doughnut, the puddle, and the phone.

That computation of C amounts to a model. The prediction engine builds a description in which Kevin has a property—call it substance C. The substance is invisible. It cannot be directly observed. It doesn't block or reflect light. It's generated inside Kevin and flows outward. It has a bias toward flowing out of the eyes along straight lines, although it does not necessarily take that path, since it can also sometimes reach out in nonvisual directions. It makes contact with specific objects in the environment. It can be partitioned among objects, but only in a competitive manner, such that as it becomes focused more on one object, it's relatively withdrawn from others. Like water from a hose, the more Kevin sprays at one object, the less he has to spare for another.

Substance C acts a lot like a classical fluid.[10] It flows from a source. Its total volume is roughly conserved—in the sense that more of it flowing here means less of it flowing there. It's also slightly viscous—it changes direction with a certain degree of sluggishness. At the same time that it acts like an invisible, viscous fluid, it also has an energy-like or will-like property in the sense that it empowers the agent. It does not directly galvanize Kevin to act, nor does it determine the specific action. Instead, its presence acts like an energizing field that powers Kevin to make a behavioral choice.

Substance C is a construct of the prediction engine. It doesn't really exist as such. It's a proxy for the very real, very complicated, neuronal processing occurring inside Kevin's head. Eighty billion or so neurons inside his brain control his actions, and the prediction engine attributes to him something much simpler, something cartoonish and very similar to metaphysical consciousness.

Substance C is a simplified, model version of cortical attention. It's an attention schema.

One can see slightly creepy practical applications for this type of artificial prediction engine. For example, it could be installed in stores to predict the behavior of people as they shop. It could be used for security or crowd management, monitoring people's attention in a way that's 10 steps more sophisticated than merely tracking who is present and where each person happens to be looking. It could be used in video games to make nonplayer characters better at predicting the behavior of real players. Attributing an invisible, metaphysical force of consciousness to agents is not an exalted form of human poetry; it's a simple, programmable trick, useful for behavioral prediction.

Our Predictor 5000 might not have a high success rate. It's a tricky thing to predict real human behavior with any accuracy. Kevin might simply walk around the room doing nothing special, muttering to himself, a behavior stream that's not very amenable to prediction. His reactions may be chaotic or determined by hidden factors we know nothing about. I don't think we humans are astonishingly good at predicting each other's moment-by-moment behavior, and I don't suppose artificial intelligence would be good at it either. But the machine doesn't have to be good in an absolute sense. If its predictions are *relatively* good, better than chance, they would still confer a useful advantage—the same advantage we humans get from our imperfect social predictions.

Let's extend our machine to handle an even more difficult situation. Kevin can direct his attention to an idea or an emotion just as much as to a doughnut or a cell phone. Adding in abstract targets of attention rather opens up the task. Imagine our Predictor 5000 machine using its microphone to listen in on Kevin's conversation as he talks on the phone. Its job is to reconstruct Kevin's fluctuating attention toward the ideas and the topics of the conversation and to predict how Kevin might respond next. The telephone task is obviously much harder than the video task, because the information the machine must gather is both greatly reduced and much more abstract. The words Kevin says and his tone of voice must convey everything. But the underlying principles are the same. The machine identifies items, in this case abstract ideas, that might attract Kevin's attention; the machine models a substance C inside Kevin, an awareness essence, that is divided among those items; and the machine uses that model to predict Kevin's verbal behavior.

I suggest that humans contain prediction engines of exactly this sort. We constantly attribute to each other a subtle substance C, a consciousness of things, a kind of force or essence that is invisible and can flow like a fluid. It emanates from a person, with a bias toward flowing out of the eyes along a straight line. It makes contact with objects in the environment. It has an energy- or will-like quality in the sense that it empowers people to make behavioral choices and thereby act on the world. We don't necessarily realize we're building this quirky model of attention. It comes automatically, giving us the curious impression that we can perceive other people's consciousness emanating from them. And we do all of this because evolution found a pragmatic solution to an important problem—predicting the behavior of others.

WE DID A fun experiment in my lab recently.[11] Volunteers looked at a computer image of a paper tube standing upright on a table. The participants were asked to imagine the tube being gradually tilted and to judge the critical angle at which it would probably fall over. With the arrow keys on a keyboard, the participants marked out their estimate of the angle, trial after trial, with paper tubes of various heights and widths. Sometimes they were asked to imagine a leftward tilt, sometimes a rightward tilt. People were reasonably good at the task, estimating physically plausible angles.

At the same time, on every test trial, we included a profile of a face on the display screen. We didn't explain the face to the participants; it was just present, either on the far left or far right of the screen, looking directly at the paper tube. When we asked the participants afterward why the face was in the experiment, they offered a variety of explanations, but never guessed the actual purpose of the experiment. Most of them thought the face had no effect on their performance. And yet it did subtly affect their tilt judgments. It was as if participants perceived a beam of energy coming out of the face's eye, pushing on the paper tube, influencing its critical tilt angle. When the tube was tilting toward the face, the eye beam seemed to prop it up, and people judged that it could be tilted further before falling over. When the tube was tilting away from the face, the eye beam seemed to give it an extra nudge, and people judged that it would fall over sooner, at a shallower angle. Yet when the face in the picture was blindfolded, the effect went away. The estimated tube tilt was the same whether it was angled toward or away from the blindfolded face, as though the blindfold cut off the eye beam and its power.

We were even able to compute the amount of fictitious force exerted by the beam, given the effect it had on toppling a light paper tube. The magnitude was very small, just enough to subtly nudge over the tube: about one-hundredth of a Newton. Just for reference, that's a little less than the force exerted by one raisin resting on your hand in normal Earth gravity. The subtlety of the force is expected, of course. If people perceived a strong force emanating from other people's eyes, like a wind that can knock over a trash can, then the conflict between perception and reality would become obvious, and the misperception would be a major survival liability. Instead, the fictitious force is so subtle that it is barely measurable.

The people we tested didn't realize the bias they were showing. In our questionnaire afterward, they insisted that their tilt judgments had nothing to do with the face. Just for good measure, we then asked them how they thought vision worked: did it involve something coming out of the eyes or something going in? About 5 percent had a confused under-standing and indicated that something comes out of the eyes. The rest correctly indicated that vision works by light entering the eyes. And yet, at an implicit level, they all seemed to be treating an open eye as though an invisible substance flowed out of it and interacted with the physical world. They couldn't help it. In my interpretation, we were tapping into substance C. We were observing a simplified model of attention at work.

The idea that visual attention might emanate from the eyes and phys-ically affect objects in the outside world is hardly new. It's one of the most persistent beliefs in folk science. It is called the extramission theory of vision and dates back at least to the ancient Greek philosophers.[12] The theory has had a long run, with illustrious proponents such as Plato and later the Roman physician Galen. In the ninth century AD, the Arab scientist Ibn al-Haytham finally worked out the correct laws of optics and declared that extramission was wrong. Light enters the eye in straight lines to form an image.

Despite that definitive scientific answer, a folk belief in eye beams persisted. A thousand years after al-Haytham, a belief in an "evil eye" is still common across many cultures, along with a lucrative trade in amulets that can protect you from it.[13] In our culture, Superman has X-ray vision, which somehow shoots out of his eyeballs and burns things. Almost everyone has had the spooky feeling of being stared at from behind, as if a beam of energy is touching the back of your neck. In 1898, the psychologist Edward Titchner thought the belief was widespread enough to be worth testing.[14] In controlled experiments, he found that people can't directly feel each other's stares, no matter how compelling the psychological impression may be.

A belief in eye beams does seem to be psychologically seductive. When I was about 5 years old, I was sitting on our porch steps with my father, looking up at the stars. He asked me how I was able to see such distant objects. In retrospect, I think he wanted to make a point about ancient light traveling for millions of years, but I'm afraid I derailed the conversation. I started to explain that when you look up, a vision thing comes out of your eyes and goes up into the sky. I could see the scientist part of him cringing. He lost no time in kindly explaining the basics of optics to me. Yes, I remember the exact moment when I switched over from an extramission to an intromission theory of vision. Maybe that was a defining moment in my life as a scientist, though it turns out that everyone goes through that transition. The belief that vision involves something beaming out of the eyes is so intuitive—built so deeply into the way we understand active vision— that it is the default belief among children, as the psychologist Jean Piaget discovered.[15] We're only later taught the scientifically correct account.

A series of studies from the 1990s suggested that most U.S. college students believed the incorrect, extramission theory.[16] In those studies, as much as 60 percent of college students insisted that vision was caused by something flowing out of the eye rather than flowing in. I should add, however, that I am skeptical of that extreme claim. I don't know if the form of the question influenced the response, or if science education has improved radically in the past 30 years, or if maybe the participants were simply trolling the experimenters. Our own findings contradict that large proportion of incorrect beliefs, and we studied a fairly broad demographic range across levels of education and income. As a serious intellectual belief, the eye beam theory of vision seems to be rare these days, and most adults understand basic optics.

And yet, despite that scientific understanding, we still seem to cling to the extramission way of thinking. To me, the most interesting finding in our paper tube study was the startling contrast between people's intellectual knowledge and their unexamined intuition. It didn't matter if they knew better; at an implicit level, they still treated vision as something emanating out of a person's eyes. I suspect that the intuition probably extends beyond visual attention to any kind of attention. Our culture is filled with naive beliefs about how intensely focused attention can reach out and physically touch people or nudge objects. I wonder how many kids, after watching *Star Wars* for the first time, tried focusing their minds on a pencil to see if it would move. I know I tried. The classic eye beams are, in my view, more like *mind* beams that often come out of the eyes, but are not necessarily limited to vision.

The reason why these beliefs have so much cultural traction may be that they tap into a deep, automatic, implicit model with which everyone is born and that evolved over millions of years. That model helps us keep track of other people's attention in an efficient, schematic way so that we can better predict their behavior. Even when we know better intellectually, we can't help generating that intuition. We can't help implicitly taking into account other people's attention beams.

I think that intuition is also the reason why consciousness is so difficult to study scientifically. Many theories of consciousness are trapped in a nonscientific, essentially mystical assumption. If we are trying to understand how the human brain generates an ethereal substance C, then we will never come to any scientific understanding. People look inward and access models that didn't evolve to be scientifically accurate descriptions of the physical world. Those models have been shaped and honed over the last 300 million years of evolution or more, optimized to be useful in specific, pragmatic ways. They are schematized and stripped of unnecessary detail. We're tempted to take the contents of those models literally and then launch a scientific search for eye beams or some inner, invisible energy or subjective essence. We try to understand how the brain might generate substance C. I doubt we'll find an engineering solution for how to build a magic mental field. There is, however, excellent engineering sense in building a behavioral prediction machine that incorporates exactly that kind of simplified construct at the heart of its computations.

THE ATTENTION SCHEMA THEORY, along with other mechanistic theories, is sometimes accused of devaluing consciousness, even dismissing it as an illusion (and I'll have more to say about the proposal of "illusionism" in a later chapter). But in the attention schema theory, because consciousness can be understood mechanistically, its specific, pragmatic uses can also be understood. Rather than dismissing consciousness, the theory places it at the center of our abilities. It is an ancient, highly simplified, internal model, honed by evolution to serve two main useful functions, as I have described over the last several chapters. Its first function may have been as a self model, to monitor, make predictions about, and help control one's own attention. The second function may have been as a catalyst for social cognition, allowing us to model the attentional states of others and thus predict their behavior. My point is not just that consciousness can be understood scientifically or that it may ultimately be built by engineers, but that it is a tool of extraordinary power and practical significance.

CHAPTER 6

Yoda and Darth: How Can We Find Consciousness in the Brain?

THE FIRST QUESTION anyone asks about consciousness and the human brain is Where? In which part of the brain is consciousness located?

Knowing where a function like consciousness is located doesn't explain what it is or how it comes about, of course. Suppose you knew nothing about how memory is stored in a computer. A technician could open up your computer, point to an object inside, and say, "That's the memory chip." With that information, you would know little more than you did before, but at least you would have a start. You'd know what object to probe for answers. Hopefully, locating consciousness in the brain will be useful in the same way.

The most commonly suggested brain structure for consciousness is the cerebral cortex, the part of the brain that expanded most in human evolution.[1] A second common suggestion is the thalamus.[2] As I've noted before, the thalamus is closely connected to the cortex, and information constantly resonates between the two, each part of the thalamus communicating mainly with a specific part of the cortex.[3]

A third suggestion is a mysterious brain structure called the claustrum, a thin sheet of cells that lies just beneath the cortex on the two sides of the brain, near the ears.[4] In 1987, when I was a naive, wide-eyed undergrad and first started to work in a neuroscience lab, I asked my adviser if I could study the claustrum, specifically because nobody seemed to know anything about it. He let me try, but the claustrum is such a thin wisp of a brain area that it was hard to get any reliable signal out of it, and I moved on to other projects. I still have a personal fondness for the claustrum and always read about it when it shows up in the science news. It's connected to the cortex in a processing loop not all that different from the thalamus-cortex loop. But even though the connections have been traced, the function of the claustrum remains unknown. The same could be said of a lot of brain structures. It's one of the delights of neuroscience that most of the brain is an uncharted mystery.

All three of these suggested brain structures are so closely connected to each other that they form one system. It would be fair to say that human consciousness is probably a function of the cortex*, where the asterisk means "plus extra structures like the thalamus and claustrum, without which the cortex doesn't function." Because the cortex is spread out conveniently on the surface of the brain, it's much easier to study than other structures and is where the main experimental effort goes. The search now is to figure out whether all parts of the cortex contribute to consciousness, or if some regions are more important than others. Whichever parts of the cortex contribute most to consciousness, the connected parts of the thalamus, the claustrum, and many other brain structures are also presumably involved.

Neuroscience is sometimes guilty of what has been termed blobology—painting a red blob on the surface of the brain, assigning a function to it, and acting as though insight has been gained. The cortex is traditionally carved up into areas about the size of postage stamps, each with its own distinct properties. But as the cortex is better understood, scientists have increasingly realized that it is made up of distributed networks, not isolated areas. Related areas open up functional connections with each other, almost like countries opening up diplomatic channels, and those alliances between areas shift and change, depending on the state of mind of the person and the specific mental task that is being performed.

In this chapter, I'll describe some of the experimental attempts to figure out which specific networks in the cortex are responsible for consciousness. I won't pretend to cover all or even most of the excellent experimental work that has been done. I've selected only a few examples in order to give a sense of the deep, sometimes hidden conceptual difficulties inherent in studying consciousness.

THE NEUROSCIENTIST Sabine Kastner works one floor down from me at Princeton University. She is a world-renowned expert on visual attention. Given our collaborations, it's not surprising that I developed a theory of consciousness centered around the current science of attention. Dr. Kastner is, by the way, not just a collaborator—she's my wife. We've had many a conversation about consciousness and the brain over the dinner table. The decorative centerpiece on the table is a three-dimensional printed model of our 11-year-old son's brain, based on a brain scan. My wife and I sometimes enthuse about its convolutions, while my son rolls his eyes and tries to steer the conversation back to his own interests (late Cretaceous fauna).

A few years back, as a part of her standard colloquium talk, my wife used to treat her audiences to a demonstration. She would hand out 3-D glasses, which separate the images seen by the right and left eyes. Then she would show a picture of Yoda rigged to enter your right eye and Darth Vader tuned to enter the left. From the perspective of the audience, you don't see both Yoda and Vader at the same time. What happens is something called binocular rivalry. First you see Yoda. Then, after a few seconds, Yoda starts to look patchy, fades away, and Darth appears. A few seconds later, Darth fades and Yoda triumphs. It goes back and forth in a yin and yang of the Force, binocular style.

Binocular rivalry was one of the first tools that researchers used to try to locate consciousness in the brain, and it is still one of the most popular.[5]

Suppose you put a person in an MRI scanner to measure brain activity, and you find a patch of visual cortex where the neurons are specifically responsive to Yoda. Maybe they selectively process green images. When you show Yoda by himself, the neurons there become active. When you show Darth Vader, the neurons fall silent. Now, suppose you show the dual, competing images—Yoda in one eye, Darth in the other—and measure the activity of those green-selective neurons.

Because Yoda is continuously present in the right eye, the neurons might be continuously active. If so, those neurons are probably not related to visual consciousness, but instead linked to a raw processing of the image as it is actually displayed. But suppose, instead, those neurons are active whenever the Yoda image has risen into the person's consciousness and then fall silent when the person reports that Yoda has faded and Darth has entered visual consciousness. In that case, the activity of the neurons correlates with a consciousness of Yoda. Maybe those neurons, as they become active and process Yoda, *cause* the person to be conscious of Yoda. By implication, there must be another set of neurons that, when active, causes the person to become conscious of Darth Vader. At least on the surface, the experiment seems to make perfect sense.

This method was first used in the 1990s in research on the monkey brain[6] and is still being used in the human brain to find the parts of the visual system that correlate with, and perhaps generate, consciousness.[7] Originally, researchers assumed that cortical areas at the bottom of the processing hierarchy, where information first enters the system, such as area V1, would respond to the images from both eyes in a steady manner, indicating no relationship to consciousness. Those areas, after all, seem like they would process information at too simple a level to create consciousness. In contrast, cortical areas at the top of the hierarchy might respond in a way that switches back and forth between the two images, tracking the way consciousness switches between them. These areas process visual information in a richer, more complex, more holistic way and therefore should be the ones to generate visual consciousness. Somewhere between those two extremes, a switchover was expected to occur. As visual information makes its way up the processing hierarchy, becoming more elaborated and more complex, it should cross a threshold into consciousness.

This simple result never did materialize. Instead, the effects are spread out over the entire visual system.[8] Not just the cortical system, but even an earlier processing stage, the region of the thalamus that connects the eye to the cortex, shows some partial correlation with conscious perception.[9] Its activity adjusts back and forth between the right and left eyes as the person becomes conscious of one or the other image. The higher up the cortical hierarchy we look, the more the neurons tend to correlate with conscious visual perception, but there does not seem to be any simple divide between conscious parts and nonconscious parts.

It's tempting to conclude that consciousness emerges gradually across the visual system, accumulating as information ascends the processing hierarchy. In that view, the brain does not contain a special consciousness-generating spot. Instead, consciousness is a property of the whole system. Could this interpretation be correct, or does the experiment on binocular rivalry, despite its elegance, have some hidden conceptual demons?

SUPPOSE OUR TASK is to understand how people see color. In some ways, color is like consciousness. We don't think of color as having any intrinsic substance or weight, but it nonetheless exists and can be attached to objects in the world around us. In what follows, I'll take a perspective on color that is wildly wrong—but the logic is similar to the logic people sometimes apply to consciousness.

Imagine that you're looking at an apple. A traditional question in neuroscience would be, Where in the brain does the apple information go to generate a conscious visual experience? Let's ask a parallel question: Where in the brain does the apple information go to generate redness?

The visual system processes the shape of objects. V1, the cortical area at the lowest rung of the processing hierarchy, handles simple aspects of shape, such as the tilt of a small line segment at the boundary of a shape. Farther up the hierarchy, in higher-order cortical areas, more complex and integrated aspects of shape are put together. Let's formulate a theory of color—I know it's wrong, but bear with me—in which color emerges from the brain's processing of shape. In that theory, when visual information first enters the system, it is without color, encoded only as black and white. Now suppose that somewhere in the processing sequence, color begins to emerge as the shape information is processed in a more complex manner. By the time the information reaches the highest levels of the visual hierarchy, color is fully generated from the shape information. In this strange theory, color is an essence generated by the brain, a by-product of how shape information is handled. Let's call it the emergent color theory.

Having decided on this incorrect theory of how color works, our next step is to figure out which parts of the cortex are the color generators. To do so, we'll use binocular rivalry on a human subject while measuring her brain activity in an MRI scanner. We'll use exactly the same logic as for the consciousness experiment I described before. For our two images, we'll use a red square and a red circle. The square is projected to the right eye, the circle to the left eye. Our human guinea pig now has an experience of rivalry—she sees a red square for a few seconds, then the square fades and she sees a red circle, and so on, back and forth. The property of redness is attached first to the square, then to the circle, in alternation. At one moment the square information must have reached the crucial parts of the brain where it generates an essence of redness. The next moment, the circle information must have surged up into the correct brain area to generate redness. Where are those brain areas?

Suppose we find a region in the cortex that is active when our subject reports seeing the red square, but falls silent when she reports seeing the red circle. That brain area must be a color generator. Its neurons process information about squareness, and when they are excited, their activity must also cause redness to emerge from squareness. Presumably, another cluster of neurons, active when the subject reports seeing a red circle, must serve a similar redness-generating function for the circle.

When we do the experiment, we discover that the telltale signs of rivalry between the square and the circle are distributed across the whole of the visual cortex, from the input end to the highest levels of the processing hierarchy. Shape information, wherever it happens to be in the system, seems to create color. From this result, we conclude that color is a distributed, holistic property that cannot be pinned to any one part of the brain. It emerges collectively from the whole system.

I realize the absurdity of the emergent color theory and the weakness of my proposed binocular rivalry experiment. I am not describing them here to mock previous approaches to consciousness, but to clarify how easy it is to fall into a flawed line of thought. After all, before color was scientifically understood, the emergent color theory might have seemed plausible. Stranger theories have been proposed. Today we know more about color, but it is all too easy to fall for an equivalent emergent consciousness theory. Here are a few of the logical problems with this line of thinking.

First, color is not an emergent property. In a sense, yes, seeing in color "emerges" from networks of neurons. The same can be said of any computed property in the brain. But for our test subject to report that the square is red, something in her brain must have computed specific information about the color red. You can build a visual system that is as spectacularly complicated and integrated and holistic as you like; it can process shape up and down and every which way; but I can guarantee you that if you leave out the part that specifically computes information about color, the visual system won't be sensitive to color. Color will be irrelevant to it. The idea that if the squareness can be intensely processed or made extremely complicated or processed in a special high-level area of the brain it will start generating color is not correct. If color is a ghostly essence generated by shape processing, if there is no specific color information being computed like any other feature of the stimulus, then how is the person able to verbally report the color? After all, to make a claim, the brain must have the requisite information on which the claim is based. To understand how people see a red square, we shouldn't put blinders on and look only at brain areas that process shape information about the square, hoping to find out which areas also generate color. Instead, we should be looking for the circuits where information about color is computed.

In the same way, to understand visual consciousness, we shouldn't be looking only at brain areas where visual information is processed, hoping to find out which of them generates an essence of consciousness. We should be looking for the networks that construct specific information about consciousness.

The second flaw in our fictitious experiment is that both stimuli have color and are even the *same* color. Whether our test subject perceives a red square or a red circle, she's still processing red. Any part of the visual system that processes color will respond in the same way to both of those conditions; it won't turn on and off as the perception flips from square to circle. If there are cortical areas that specialize in color information—and indeed there seem to be many cortical hot spots for processing color[10]— then our experiment would specifically fail to find them. Similarly, when studying consciousness, whether the test subject is aware of the left-eye image or the right-eye image, she's equally aware of an image in both cases. Any brain region that is computing information about the presence of consciousness should be equally active in both cases. It should not switch on and off depending on whether one or the other image is dominant. The experiment would specifically miss areas of the brain computing the construct of consciousness.

The third flaw in the experiment is a failure to understand where binocular rivalry comes from. Binocular rivalry is caused by competition.[11] Signals from the two eyes begin to interact almost as soon as they enter the brain. That interaction grows as the information passes deeper into the visual system. If the images arriving through the two eyes are sufficiently different, as in the case of Yoda and Darth or a square and a circle, then the system cannot usefully fuse them, and instead the two sets of signals directly inhibit each other.

The right-eye signal and the left-eye signal are like equally matched wrestlers. First one comes out on top, then the other. Neither can win permanently. Whichever signal has temporarily won dominates the cortical system, submerging the other signal. When you find rivalry effects in the thalamus or in V1 or at any other stage of the visual hierarchy, none of those findings indicate that consciousness itself is necessarily present or computed or processed in a particular brain area. The results show, instead, that the two visual signals, one from each eye, are in a constant, running competition throughout the entire visual system.[12] Whichever image wins the system-wide competition is eventually attached to consciousness, but we are no wiser about what brain system constructs consciousness or how it is attached to visual information.

Using binocular rivalry to study consciousness is a good illustration of how an experiment can seem elegant at first, but when you examine the deeper assumptions, it can turn into a scientific quagmire. The mistakes in logic may be obvious when dealing with how people see color, but the same mistakes become astonishingly easy to slip into, even intuitively comfortable, on the topic of consciousness. Binocular rivalry is still a wonderful phenomenon worth studying at the level of neurons and brain areas. Many scientists, my wife included, continue to explore it for its own sake and for what it says about visual competition. But it turns out not to be a straightforward tool for finding consciousness in the brain. We need another experiment.

I AM GOING to suggest an experimental approach that is a lot less clever than the binocular rivalry experiment. It's blunt and approximate, but it's pragmatic. Let's start by searching for color-processing networks in the brain and then extend the logic to finding consciousness-processing networks.

If I wanted to find the color-processing hot spots in the cortex, I would place our test subject in an MRI machine and show her a series of black and white pictures. Every now and then, I would throw in a colored one. If a part of her visual cortex lights up more when the colored picture appears, then I would know to study that brain area further.

The experiment is encouragingly simple. As a general rule, the fewer working parts an experiment has, the better its chances of success. But at the same time, this proposed experiment is not perfect. For one thing, black and white are colors, too, processed much like any other color. The experiment compares pictures that have *less* color range with pictures that have *more* color range, and hopefully the difference is enough to light up the color-processing areas. A second pitfall is that the experiment might light up a lot of extraneous brain areas that have nothing to do with color. A brain area might help to control alertness, for example, lighting up when a colorful image pops up unexpectedly after a minute of boring black and white images. But despite all its imperfections, I'd trust this method to get at least an initial purchase on the question.

The method has been used before and does an excellent job of picking out the major color networks.[13] By itself, the experiment is only a start, but other methods have converged on the same brain regions. For example, if you test people who have suffered stroke damage to those brain areas, you find that those patients have lost their color perception.[14] They see the world in gray and even lose the capacity to remember or fully understand what color is.

With many simple experiments coming at the problem from different angles, the findings become more and more secure. In science, in my experience, it's better not to obsess over creating that single perfect experiment that is exquisitely clever and tied up in a bow. The cleverness can get in the way. Successful science is usually more like a series of approximations and slowly growing confidence.

Now let's apply the same imperfect, but pragmatic approach to consciousness.

Suppose I sit our test subject in front of a computer screen. She stares at a spot at the center of the screen. Suddenly, and very briefly, she sees a picture of a face. It's dim and lasts for a mere twentieth of a second. After the face disappears, a random set of colored squares covers the screen. That visual mask, as it is called, interrupts the brain's processing of the face and makes it that much harder to see. The face teeters at the edge of the participant's awareness. Every few seconds, another face is presented under the same difficult conditions. Sometimes our test subject says, yes, she saw the face that time (let's call these circumstances condition A), and sometimes she says, no, she didn't see a face (let's call this condition B).

Here we have a simple, blunt, if admittedly imperfect tool for finding consciousness in the brain. In condition A, the participant is visually conscious of the face and the colorful mask that covers the screen after the face. In condition B, she's conscious of only the mask, not the face. In a sense, she has more visual consciousness, or is conscious of more varied visual images, in condition A. If we find a brain area that's more active in condition A, then maybe that brain area participates in visual consciousness. It would be a useful first pass, anyway. The experiment might pick up extraneous brain areas unrelated to visual consciousness, but we would at least have some initial purchase on the question, and then we could move on to other experiments, such as studying stroke patients who have damage to those same brain areas.

This particular experiment with a face may have never been done exactly as I've described it, but it belongs to a general category of experiment on visual consciousness that has been tried many times.[15] In condition A, the participant is aware of a specific visual image. In condition B, the same or a similar image is presented, but the participant is unaware of it.

You might think that in condition A, when the face reaches consciousness, the visual system must be processing information about the face, whereas in condition B, when the face fails to reach consciousness, the visual system must obviously fail to react to it. But that pattern is not typically what happens. Areas of the brain known to process shape, color, texture, and other visual features are buzzing away in reaction to the image, whether or not the person consciously sees it. There are, however, some notable differences between the conscious and nonconscious conditions. The activity in the visual system tends to be more sustained and less variable in the conscious case than in the nonconscious case.[16] This difference in activity fits the general idea that when visual signals are boosted and stabilized by attention, they are more likely to reach consciousness. The biggest difference, however, appears in a specific set of brain areas in the parietal and frontal lobes.[17] These areas are robustly activated by the visual image in condition A, when the person says she is conscious of it, but are much less activated by the image, if at all, in condition B, when the person is not conscious of it.

This result has been repeated many times in many variations, which gives me some confidence in the findings. One of my own recent experiments obtained a nearly identical result.[18] When you are shown something and become conscious of it, the parietal-frontal networks react. When you are shown something and your visual system processes it, but you do not become conscious of it, those same networks tend not to react. The simplest interpretation is that the parietal-frontal networks, or some subsystem within them, may build consciousness.

ONE PARTICULAR PART of the parietal-frontal complex is often suggested as a crucial hot spot for consciousness. The prefrontal cortex—the front-most part of the cortex just behind your forehead—is still poorly understood, but what is known about it suggests that it is an extremely high-level processor of information. The standout property of the prefrontal cortex is its flexibility.[19] It seems to be engaged in almost any kind of task you choose to do. If you are counting dots, the neurons in your prefrontal cortex track the numbers.[20] If you are monitoring the location of dots on a computer screen, the same neurons seem to be tuned to location.[21] In a famous experiment by Earl Miller at MIT, monkeys were trained to categorize ambiguous animal drawings into either cats or dogs.[22] The neurons in the prefrontal cortex soon turned into cat and dog detectors, some neurons responding when the monkey indicated that he saw a cat, others responding when he saw a dog. The sheer arbitrariness of the task—the monkeys had never seen a cat or dog in real life—showed that prefrontal neurons can take on any properties, depending on the task at hand.

Neuroscientists sometimes refer to the prefrontal cortex, especially a large subregion called the dorsolateral prefrontal cortex, as the "working memory" part of the brain.[23] Data from other brain areas can evidently be uploaded to that prefrontal area, where the information is temporarily maintained and manipulated to serve the needs of the moment.

Given these properties, the prefrontal cortex seems like a natural candidate for the theater of consciousness.[24] It collects information through its widespread connections to the rest of the cortex, like a spider at the center of a web.

And yet this solution, as tempting as it may be, does not satisfy me. Too many nagging questions remain. For example, when people suffer damage to the prefrontal cortex, why don't they lose their experience of consciousness? Although they have difficulties planning for the future and switching easily from task to task, they do not generally lose their sense of consciousness.[25] At least, that is not a standout, classic property of prefrontal damage.

But I have an even more basic question about the prefrontal cortex hypothesis: when information reaches that brain area, why would it generate a conscious feeling anyway? As long as neuroscientists are looking for a theater of consciousness—a central brain region where information converges and generates consciousness—then I think the science is stuck.

It's the same problem I outlined earlier. Where does the shape information go to generate color? Nowhere. To understand color, we need to look for where color information is computed. Where does visual information go to generate a conscious experience? Nowhere. To understand consciousness, we need to look for a system in the brain that computes *information about consciousness*—about its properties and consequences. The information about consciousness, computed in one system, is linked to the visual information about the object you're looking at, computed in a different system. The two packets of information, joined together in a temporary alliance across brain networks, form a larger internal model that says, in effect, "Here is my conscious experience, there is an apple, and right now the one is attached to the other."

I am happy to accept the prefrontal cortex as a part of the process. Maybe it acts as a mental forum just as scientists suspect, as an active gathering of goals, thoughts, and observations, helping to organize our behavior. The idea is powerful and fits the data. But I suggest that the gathering of signals from around the brain must contain, somewhere within it, information about consciousness, or else we would never make any claims about that property. We already know which networks in the brain compute visual information about an apple. The visual system has been intensively studied for nearly a century. But where are the networks that compute information about consciousness?

MY COLLEAGUES IN neuroscience know how vague I am being when I talk about networks that stretch from the parietal lobe to the frontal lobe. At least a dozen have been studied in detail.[26] They include, just to name a few, the dorsal attention network, the ventral attention network, the salience network, the control network, the theory-of-mind network, and the default mode network. And there are more, in some cases with extremely specific functions. One network is related to the control of eye movements and probably overlaps with the control of attention.[27] Another network is related to the control of reaching.[28] Yet another, to the shaping of the hand in order to grasp objects.[29] In my own lab, before my work turned to the topic of consciousness, I studied a parietal-frontal network that processes a cushion of space near the body, the peripersonal space, and helps to coordinate defensive actions toward looming objects.[30] Another network is believed to play a role in counting and mathematical reasoning.[31] The diversity is incredible. Adding to the confusion, many of these networks have unclear borders and are recruited in many different, overlapping tasks. Where in this chaos of networks and areas might consciousness be computed?

My current best guess is that computations specific to building the construct of consciousness are especially emphasized in an area of the cortex called the temporo-parietal junction, or TPJ.[32] The brain has two TPJs, one on each side, more or less just above the ears and on the surface of the cortex. The TPJ on the right side of the brain may be more extensive or better developed than the one on the left, although I think the exact nature of the asymmetry is not yet clear. Within each hemisphere of the brain, the TPJ is itself divided into subregions with different properties, and the uppermost division, the one that lies mainly in the parietal lobe rather than the temporal lobe, is of particular interest to me. In my lab, we call this region the dorsal TPJ, or TPJd (see Figure 6.1). It is also sometimes called the inferior parietal lobe. The TPJd is positioned at the crossroads of the control network, the ventral attention network,

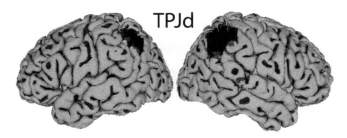

Figure 6.1 The upper, or dorsal, segment of the temporo-parietal junction (TPJd) on the left and right sides of the human cerebral cortex, determined by a convergence of two methods: first, a technique called independent component analysis, and second, the pattern of connectivity to the rest of the cortex. (Reprinted from p. 9435 of K. M. Igelström, T. W. Webb, and M. S. A. Graziano, "Neural Processes in the Human Temporoparietal Cortex Separated by Localized Independent Component Analysis," *Journal of Neuroscience* 35 (2015): 9432–45.)

the salience network, and the theory-of-mind network.[33] It is one of the most massively connected hubs of information in the brain, rivaling even the prefrontal cortex for connectivity. If it contributes to computations about consciousness, it probably does not work alone, since complex computations are likely to depend on distributed networks rather than circumscribed areas. It also probably has many functions well beyond computations about consciousness. In what follows, I will refer to the TPJ in general, rather than to the specific TPJd subregion, because we lack enough information to rule out the other components of the TPJ. We suspect, however, that the TPJd may be playing an emphasized or special role in computations about consciousness.

I want to be clear about what I'm suggesting. In my proposal, the TPJ is not, itself, conscious of anything, nor does it generate a conscious experience. It is not a homunculus—a little person hiding in the head. Instead, it is part of a network that builds a construct, a model, that informs the brain about what consciousness is. Without that information, we would not be able to attribute consciousness to ourselves, claim to have consciousness, or even understand the question when asked about it.

I have several reasons for highlighting the TPJ. First, experiments over the past 30 years have shown that it is involved in processing the possible mind states of other people.[34] When you think about other people's beliefs, emotions, or intentions, your TPJ lights up as a central node in a theory-of-mind network. If we are looking for brain areas that may supply us with the very notion of a conscious mind, the TPJ is a natural place to look. In my lab's experiments, when people think about consciousness, whether they are answering questions about their own consciousness ("Yes, I am aware of that image") or attributing states of consciousness to others ("Yes, I think Kevin is aware of that object next to him"), the TPJ reacts.[35] It lights up as though participating in the construct of consciousness.

A second reason why I suspect that the TPJ might be involved in consciousness is its relationship to attention. Recall that in the attention schema theory, the brain constructs information about consciousness for a specific reason. The information acts as a useful attention schema, a model that describes some of the properties of attention. Any cortical network that helps to control attention ought to have access to an attention schema. Therefore, we should look to the attention-related networks—the dorsal attention network, the ventral attention network, the salience network, the control network. These networks are thought to be engines of attention, able to influence other areas of the cortex to enhance some signals at the expense of others.[36] They are natural guesses for where an attention schema might be constructed. In a range of experiments, including those from my own lab, brain activity associated with attention and brain activity associated with social cognition show a specific area of overlap in the TPJ, again especially in its upper half.[37]

One concern that I have heard about the involvement of the TPJ in consciousness is that when you become conscious of a visual image, such as the image of a face, there is no evidence that information about the face ever reaches the TPJ—or indeed any of the areas in the parietal lobe. The identity of a face, the emotional expression on the face, the fine visual details that make up the face—none of that information is thought to be processed in the parietal lobe. In a now classic theory, in the 1990s, psychologists David Milner and Mel Goodale[38] suggested that the parietal cortex is specifically *not* visually conscious, whereas other areas of the visual system, especially in the temporal lobe, *are* visually conscious.

Other scientists have argued against both the parietal and the temporal lobes as sources of visual awareness, arguing instead that consciousness is coupled to the lowest levels of processing, where visual information first enters the cortex in area V1.[39] Right now, as I look at my computer screen, I'm conscious of the precise curves and angles of the letters that I'm typing. In looking up, I can see a mass of sycamore leaves outside the window. The colors are vivid, and the finest details and textures stand out. Area V1 extracts exactly these kinds of details. Moreover, when people suffer damage to V1, they lose conscious visual perception.[40] Given these observations, consciousness might not be generated at the top of the cortical hierarchy, where cognition and higher thought are represented, but instead might sit at the bottom, where the raw details are processed.

These previous views seem to contradict my suggestion that the TPJ is involved in consciousness. The attention schema theory, however, restructures the underlying assumptions and makes all of these perspectives compatible with each other. Suppose you are visually conscious of a picture of a face. Area V1 helps construct information about detail and color, saturation and vividness. The temporal lobe helps construct information about the identity of the face. The TPJ helps construct information about consciousness. All of those components are necessary to claim that you are conscious of a face. Milner and Goodale were right that the *contents* of visual consciousness, in this case the details and identity of the face, are not usually found in the parietal areas. But without the TPJ, there might be no construct of consciousness to attach to the contents.

A TRADITIONAL WAY to think about the visual system is like an assembly line. Packets of information enter through the eye and then pass from one stage of processing to the next, ascending a hierarchy, becoming ever more fully elaborated. Once the product reaches the highest stage, its construction is complete, and the previous stages, having performed their various jobs, are now obsolete. The highest level of the cortical hierarchy can then take that fully assembled information and turn it into actions or speech—or perhaps consciousness.

A more apt analogy might be a sheaf of blueprints for a house. Each page describes the building in a different way, at a different level of detail. One page contains the layout of the walls. Another page shows the electrical system. Yet another page shows the plumbing, and another shows the fixed cabinets. There is even a page that details the suggested furnishings. The pages form a natural hierarchy, since you can't draw one page without having worked out the details from the prior pages. Information, in a sense, flows from one page to the next, becoming ever more processed. But at the same time, as you flip through, you are not discarding the previous pages or progressing toward a final master plan, a single page with all the important information assembled on it. Instead, every page contains something useful, and if you lost any one page, you wouldn't be able to build the house. The final action, building the house, springs from each and every page of the book, not just from the final page with the most extensively processed information.

In the same way, the visual cortex has a hierarchy—probably many interacting hierarchies.[41] But each higher stage does not replace the previous ones. There does not seem to be any final stage into which everything important arrives at the end of its processing journey, to enter consciousness or trigger action. Instead, each hierarchical stage provides a different analysis, dropping some kinds of information, enriching other kinds, adding to the layers on layers of useful analysis of the visual world. That accumulation of information, from the tiny visual details represented at the lower levels to the deep conceptual information represented at the higher levels, is available to influence action and speech. The output isn't driven only from the top level of the hierarchy, but from all levels at the same time.

If a person says, "That's a picture of my friend's face," the assertion relies on information computed in face-processing regions at the highest levels of the visual hierarchy, in the temporal lobe. If the person says, "My friend has a little smudge on his nose," that comment probably depends on information from lower-level visual areas, where more of the visual details are processed. If the person says, "My friend's hair is brown," that comment is based on information from color-processing regions in specific color hot spots in the visual cortex. If the person says, "My friend has a squeaky voice," that comment may draw on information retrieved from memory and replayed in some part of the auditory cortex. If the person says, "Tomorrow afternoon I'll pay him back the twenty bucks I owe," that comment might depend on information constructed partly in the planning-related prefrontal cortex.

And if the person says, "I have a *subjective experience* of all that other content. I have a mysterious subjective self, a consciousness," that claim to consciousness may depend on information constructed in another specialized cortical network, one that may intersect the TPJ.

I do not mean that the cortex is a set of separate modules, each limited to one domain of information and isolated from the others. Instead, although specializing in different kinds of information, the cortical areas interact and influence each other in a massive, resonating network. Pluck one thread in that network—add new visual information at the V1 end or add a new thought constructed at the prefrontal cortex end—and it reverberates through the whole network, causing ever-changing patterns of activity. Packets of information are boosted by attention or sink back down into the background noise. Cortical areas are constantly forming alliances, opening up temporary lines of communication, joining their information into larger structures, and then decoupling to form alliances elsewhere.

In this framework, we don't need to pick whether consciousness is most closely associated with the bottom of the hierarchy, the top, or somewhere in between. We can be conscious of visual details computed in V1 and just as conscious of abstractions computed in the prefrontal cortex. The reason is that the notion of consciousness is defined by an information set constructed in a specialized network, and that network can form a functional alliance with a range of cortical areas.

IMAGINE AN INTELLIGENT alien who visits Earth. Suppose its brain does not have a construct of consciousness that joins the rest of its internal information. It lacks a counterpart to the human TPJ. It presumably still has self models—it can monitor its own internal processes and control its behavior—but it evolved along a different path, and its self models don't contain the notion of a subjective consciousness.

We ask the alien, "Do you see the apple?"

The alien says, "Yes, I see it."

Us: "Do you have a conscious experience of it?"

Alien: "I'm processing it visually."

Us: "But do you have something else, that extra *je ne sais quoi*, an essence of experience associated with the apple, or are you just processing information? Does it *feel* like anything, internally, to process that apple?"

Alien: "I don't know what the extra *je ne sais quoi* is. There's an apple and I'm visually processing it. What else would I do? Why would there be an extra essence? What is the essence made out of and where does it come from?"

Us: "Aha! But you use the word 'I,' so you must have a self concept."

Alien: "Yes indeed. This conversation takes place between two agents that can be termed 'you' and 'I.' I have a fairly complete data set on my physical body, my past, and my behavioral characteristics."

Us: "So you have self-awareness after all?"

Alien: "I am self-knowledgeable. What is awareness?"

Us: "It's when, in addition to merely processing information about yourself, you also have a subjective experience of it. You have an added something, a *je ne sais quoi*, a . . . oh heck, never mind."

My point is that the mind is information. I like to say that the human mind is a trillion-stranded sculpture made of data, constantly changing, beautifully complicated. If a mind lacks information about consciousness, it cannot know what that property is or attribute it to itself or to others. We humans make a big deal about consciousness only because we have a subsystem in our brains that constructs information about it. And I believe that science is homing in on where that subsystem lies in the brain and how it is connected to other systems.

A TIME-HONORED WAY to locate functions in the brain is to study the consequences of brain damage. The visual cortex, for example, was identified when damage to the back of the brain in monkeys, in the area now known as V1, caused blindness.[42] The speech centers in the human brain were first discovered when damage to a part of the frontal lobe, now known as Broca's speech area, eliminated people's ability to talk.[43] The visual motion areas were confirmed when damage to a specific part of the visual cortex resulted in motion blindness.[44] One famous patient with cortical motion blindness can't properly pour water. All she sees is a glacier hanging from the pitcher and then a mess all over the table. And damage to the color-processing regions of the cortex removes color perception.[45]

Does damage to a specific part of the cortex remove consciousness?

We're not looking for a condition that puts people to sleep. The cycle of sleep and waking is controlled by much more ancient structures in the brain stem.[46] We're looking instead for a condition in which people are technically awake, information can enter the cortical system and maybe even influence behavior, but any subjective consciousness of that information is gone.

We're also not looking for a philosophical zombie.[47] Philosophers have invented a rather specialized concept of the zombie—not the brain-eating thing that walks in a stilted manner with its arms out, but a person who looks and acts normally, even talks normally, while lacking any inner subjective experience. Scholars debate whether zombies can or do exist. The hypothetical space alien that I described earlier is a zombie. The attention schema theory allows for the possibility of zombies: maybe a zombie could be built artificially or evolve on another planet. But if the theory is correct, we could never turn a normal person into a walking, talking, philosophical zombie by taking out a part of that person's brain. We couldn't remove consciousness without leaving the person unable to function in everyday life. The human brain depends too much for its normal functioning on the construct of consciousness.

If your attention schema were surgically removed, you would suffer at least three disabilities. First, you would be unable to regulate your own attention. If you can't control your attention, then you would be fundamentally incapacitated, unable to direct actions in a sustained, purposeful way toward specific objects or goals. You would be even more impaired than the movie zombies, who, for all their stilted walking, retain a fixed attention on killing people and eating their brains. Second, you would lack the ability to build social models of other people's minds. Unable to attribute consciousness to others, your social cognition would collapse. Third, lacking the necessary information set, you would be unable to understand questions about consciousness or make coherent claims about it. A traditional, philosophical zombie can blend into the crowd, but an "attention schema zombie" is a seriously impaired individual. If you stuck it with a pin it might react, but otherwise it wouldn't do much.

In the clinical world of brain damage, there is a particularly horrible and disruptive syndrome called hemispatial neglect, which has been studied for nearly 100 years.[48] Damage to one side of the brain causes a loss of awareness of objects and events on the opposite side of the body. The critical mechanism affected in neglect is present on both sides of the brain, but damage on the right side consistently causes a greater and more permanent loss of function.[49]

Neglect is not blindness. In blindness, objects are gone from your vision, but you can still know about them. Everyone is blind to objects behind their head, for example, but you still probably have a reasonable idea of what's behind you right now. In neglect, objects simply disappear from your consciousness when they enter the bad side of space. A touch on the left side of the body is not acknowledged. A sound coming from the left is either mistakenly attributed to the right or ignored altogether. In the most profound cases, the patient doesn't even realize that a left half of space exists.

A neglect patient might shave one half of his face, try to dress one half of his body, and eat the food on one side of his plate. If you rotate the plate, he won't know where the extra food comes from. If you ask him to draw a picture, he'll draw the right side of it and not notice that he's skipped half the image. If you ask him to draw a clock, he'll typically draw a complete circle, possibly because he retains a motor memory for that circular hand motion. But then he'll squeeze the numbers, 1 through 12, into the right side of the circle, and he'll think he's done it correctly.[50]

When I was in graduate school, I met a neglect patient who visited our lab. We didn't normally do clinical research, but we knew the most basic test available. We took a sheet of paper and filled it with short, horizontal line segments scattered randomly over the page. Then we placed it in front of him and asked him to cross out every line that he saw.

I remember that he laughed and said, "This again!" He stared hard at the page and crossed out every line on the right-hand side. When we asked him if he was *sure* he was done, he stared intently and said, "You're right, I missed a few." Then he promptly crossed out a couple of more lines near the center of the page, leaving everything on the left side still untouched. We rotated the page, putting the left on the right, and he was shocked by how many lines he had failed to cross out. The test was a surprise to him every time he did it. Neglect patients know something is wrong; they just can't figure out what it is.

One of the spookiest experiments on neglect patients involved a test of visual imagination.[51] Patients were asked to imagine standing on the north side of a familiar city square and to name all the buildings they could remember. They promptly named the ones that would have been to their right. Then they were asked to imagine standing on the south side of the same city square. This time, they named the opposite set of buildings. They knew they were making a mistake, but they couldn't grasp where the mistake came from. They couldn't even imagine a left side of that city square.

And yet the information from the neglected left side is not erased altogether. In one particularly poignant experiment,[52] a neglect patient looked at a picture of a house with flames pouring out of a window on the left side. Asked to describe the picture, the patient described an ordinary house. She wasn't conscious of the flames, but she said that she didn't like the house. Something felt wrong, even if she couldn't tell why. That experiment is just one of many that show how information from the neglected side enters the brain, is processed, and can affect behavior in subtle ways. If you stick a pin in the neglected side of the body, the patient might squirm and know that something unpleasant has happened, but be unaware of any specific pain. If you throw a ball at a patient from the left side, the person might duck and not know why. Neglect is not an erasure of one half of space. It's a lack of consciousness and directed attention to that half of space.

In the context of the attention schema theory, the simplest explanation of neglect is that the mechanism for attention, including an attention schema, can be separated into at least two fields, a right and a left. On the right side of space, the patient can function, more or less, with respect to attention, action, social cognition, and the claim of subjective consciousness. On the left side, the patient becomes an attention schema zombie.

Hemispatial neglect is usually caused by strokes that damage a large area of cortex, making it difficult to pinpoint which specific cortical area is responsible for the main symptoms. By looking for the areas of overlap in the brain among many different patients, neuroscientists have narrowed down the epicenter of neglect. Damage to many brain areas can cause some degree of neglect, but by far the most severe and longest-lasting cases follow damage to the TPJ, especially the upper part that overlaps with the parietal lobe.[53] Damaging that crucial region of the brain disrupts consciousness.

Hemispatial neglect, as devastating as it is for patients, at least affects only one side of space. The patients still have a part of consciousness left. They can interact with loved ones and experience life through the other half of space. In many cases, the syndrome fades over time and the patient recovers some, if not all, of the lost ability as the brain reorganizes.

But for some stroke patients, the brain damage is so extensive that they lose all consciousness. Many of these patients are still minimally responsive. If you pinch them or puff air on their corneas, they react. They're not brain-dead—much of the brain remains intact and functioning. But they do not show any higher response. They don't appear to be conscious anymore. These are the most heartbreaking of cases, because the essence of the person is gone.

These nonconscious, or vegetative-state, patients are often scanned to determine the extent of their brain damage. The damage can be massive and can vary from patient to patient. But a pattern emerges of a core set of brain areas most frequently associated with a loss of consciousness: the parietal-frontal networks.[54] Remove those networks on both sides of the brain, and you remove consciousness. An attention schema zombie is not a philosophical flourish, a what-if, or a thought experiment. It's a real-life medical tragedy.

CHAPTER 7

The Hard Problem and Other Perspectives on Consciousness

It's easy to get the impression that consciousness is a matter of academic opinion and that the world contains too darn many opinions all jumbled in a great chaos. This impression feeds the popular idea that consciousness must be an unsolvable mystery. I am not so pessimistic. I think we're close to understanding the mechanism of consciousness and may already grasp the basic principles. One reason for my optimism is that many of the theories and opinions are not as different as they seem at first. Dig down deep enough, and intriguing commonalities emerge.

In this chapter, I'll explain how the attention schema theory might relate to seven other well-known scientific perspectives on consciousness. I will disagree with some and find common ground with others. I won't be able to give anything like a complete account of these alternative perspectives in such a short space. Rather than evaluate each theory, my emphasis will be on how it connects with the attention schema theory.

THE HARD PROBLEM AND THE META-PROBLEM

The philosopher David Chalmers coined the term *hard problem* and in doing so shaped the scholarly debate on consciousness for decades.[1] Consciousness is a hard problem because it's a private experience; it can't be confirmed from the outside. By the very nature of subjective experience, you can't push on it and measure a reaction force, put it on a scale and measure how much it weighs, or heat it and measure its combustion temperature. It's immune to science. Calling it the hard problem is really a euphemism for the *impossible* problem that science can never approach.

Chalmers has now also offered the term *meta-problem*, which refers to why we think we have a hard problem.[2] Maybe we don't. Maybe there's no fundamentally inexplicable, nonphysical essence in us. Maybe our task as scientists is to explain why people tend to believe in a hard problem in the first place.

The attention schema theory fits this second approach. When you boil down the theory to its essence, it is an explanation for how a biological machine falsely believes in a hard problem. When the machine accesses its attention schema—a simplified, cartoonish account of its own internal processes—it is informed that it contains a private, ghostly, inner property of consciousness.

The trick at the center of the theory is the power of model-based knowledge in contrast to superficial knowledge. The following example may help explain what I mean.

Suppose a child plays at make-believe. She barks, crawls on all fours, and says, "I'm a puppy!" In order to make the claim, her brain must construct the key proposition "I'm a puppy" as well as contain the information that puppies bark and walk on all fours. And yet that information exists in a larger context. Her brain contains a vast net of information, including "I'm not really a puppy," "I'm making it up to play a game," "I'm a little girl," and so on. Some of that information is present at a cognitive and linguistic level. Much of it is at a deeper, sensory or perceptual level. Her body schema is constructed automatically, beneath higher cognition, and it describes the physical layout of a human body, not a puppy body. She sees her human hands in front of her, and the visual information confirms her human identity. She remembers eating breakfast cereal with a spoon, going to school, reading a book—all human activities. The claim "I'm a puppy" is a superficial proposition that is inconsistent with her deepest internal models.

But suppose I have the science fiction tools to manipulate the information in her brain. I alter her body schema to reflect the body of a puppy. I alter the information in her visual system and her memory to make it consistent with the puppy proposition. I remove the specific cognitive information that says, "I made that up to play a game." I switch the information that says, "I'm certain this is not true," to its opposite. How would she know that she's not a puppy? Her brain is captive to the information it contains. Tautologically, it knows only what it knows. She would no longer think of her puppy identity as a hypothetical or a game. She would take it as a literal truth. There would be no reason for her to think anything else.

You could try to convince her otherwise. You could say, "But you understand English, and you can talk. Puppies can't do that. Don't you think that suggests you've mistaken your identity?"

Let's suppose she's an intellectually gifted little girl and realizes the logic of your argument. That new information will be at a superficial, cognitive level. It will conflict with her deeper internal models. She will be in a position of believing one truth about herself intuitively while entertaining a different truth intellectually.

Just so, in writing this book, I might be able to convince you that your consciousness has its basis in an attention schema. In that intellectual argument, you claim to be conscious because you contain a set of information that says so. But intuitively, you believe a different truth about yourself. When you rely on introspection, when you access that very same attention schema, it provides you with its own story. The information within it tells you that no, your consciousness is not information or mechanism or neurons—it's an ethereal essence and an inherent property dwelling inside you. Presuming I've done a good job of convincing you of my argument, you'll find yourself conflicted, with superficial, intellectual knowledge pointing you toward one understanding and deeper, internal models anchoring you to a different understanding. You will never be able to close that gap. You can't use superficial, intellectual knowledge or a few hours of thought to erase an attention schema that was shaped through millions of years of evolution and is constructed deep within your system.

To give another example of an imperfect model built deep within us, consider how we see the color white. The visual system builds a model of white as brightness in the absence of any contaminating colors. That model evolved over millions of years and is shared across many species of animal. Eventually, in the year 1671, one particularly brainy animal, Isaac Newton, figured out that this internal model is a simplification.[3] White light is a mixture of all colors, and the brain represents it in a simplified way.

We could say that the "hard problem" of the color white is, What is the special physical process that purifies white light of all its contaminants? The corresponding meta-problem is, Why do we even think this hard problem exists? Why do we think that white light is purified? We now know the answer to the meta-problem—the brain constructs a simple, pragmatic, but imperfect model—and with that, we also know that the hard problem doesn't need to be solved.

And yet, even though every educated person now understands that white light is a mixture of all colors, that knowledge doesn't change the models built into the visual system. We still see white as pure rather than as a mixture. Nobody seems to mind the contradiction. We've become used to a layer of intellectual, cognitive knowledge that conflicts with the brain's deeper, inborn models. Arguably, science is the gradual process by which the cognitive parts of our brains discover the inaccuracies in our deeper, evolutionarily built-in models of the world.

THE PHILOSOPHER François Kammerer asked an insightful question about the attention schema theory.[4] Let's suppose that the theory is correct. The brain constructs an attention schema, which represents attention. It depicts general, high-level properties, such as our ability to focus on and deeply process information. At the same time, it leaves out any depiction of the physical or mechanistic properties of attention. It doesn't specify that attention *lacks* a physical substance—it is merely silent on the topic. It is uninformative on the details of working parts, such as neurons and synapses. If our intuitions about consciousness are shaped by that internal model, then why do we have such a strong intuition that consciousness is an ethereal essence? Where do we get the intuition that consciousness is physically weightless or substance-less if that proposition is not contained in the relevant internal model?

The answer, I believe, is that people generally don't have that intuition. We don't understand consciousness as being physically substance-less. Instead, we understand it as something for which physical attributes are *irrelevant*. And those two intuitions are very different.

To understand what I mean, imagine that someone taps you on the shoulder. That touch activates receptors in the skin, which transmit information to the brain. Ultimately, your brain constructs a specific kind of internal model, a tactile model, a packet of information that describes that particular touch. The model contains information about the location of the touch, its intensity at onset, its pressure, its duration, and perhaps even the smooth or plush texture of a fingertip. It's a rich sensory representation. But it contains no information about taste. A touch on the shoulder doesn't come with a salty taste, for example. I don't mean that a touch is bland or tasteless and needs some salt; no, it doesn't lie *anywhere* on any taste dimension. It does not occupy the same information space. Now that I've mentioned the possibility, you can consider it in a superficial, cognitive sense, but you can't alter the deeper internal model. Touch perception is an inborn process and is not open to cognitive modification. You can't make a touch have a taste.

I'm pretty sure that if you could insert electrodes into the brain of a normal person and read the information encoded in the tactile system, the perceptual model for touch would *not* contain the information, "And by the way, no taste is present." It doesn't need the explicit negation. It is simply silent on taste properties. We don't intuitively understand touch to be something for which taste has been muted; instead, we understand touch to be something for which taste is *irrelevant*.

I argue that the attention schema acts the same way. It contains a rich, but nonetheless limited, set of information. It depicts general properties of attention, but not physical, mechanistic properties. Based on that internal model, we intuitively believe in a mental experience inside us that can take possession of information and prompt us to action, the way attention does, but that has no specific relationship to physicality. Physicality is irrelevant to it. That mental essence, the experience itself, is, in this sense, *metaphysical*. It isn't physically graspable, smooth, textured, rough, bumpy, heavy, light, smelly, green, or pointy. It doesn't lie anywhere on those physical dimensions any more than a touch exists on the salty dimension.

And yet in this theory, the attention schema depicts at least one physical property. It depicts attention as something with a physical location roughly inside us. With that kind of internal model, we should have an intuition about a mental essence that overlaps with the physical world, in the sense that we can point to a location and say, "It lives roughly here." It is a kind of ghost, inhabiting physical space even as it lacks other physical parameters. In this theory, the ghost in the machine, the conscious energy inside us, is an intuition that comes straight from the attention schema, with its incomplete account of attention.

And so we come back again to the hard problem and the meta-problem. The hard problem derives from assumptions that come from that deep, subsurface model, the attention schema. The attention schema theory is a meta-answer that explains why people believe in a hard problem in the first place.

ILLUSIONS AND METAPHORS

Is consciousness an illusion?

Illusionism is a relatively new and growing theoretical approach to consciousness.[5] The central idea is that we don't actually have consciousness. The experience itself, the subjective essence, is absent. Instead, we *think* we're conscious because of an illusion produced by the brain. Maybe that illusion has a specific functional advantage (to give life more aesthetic gusto, according to one proposal), or maybe it has no survival advantage and is merely an accidental consequence of how the brain processes information.[6] Someday we'll understand the mechanism behind it, and then we'll know whether it has any functional significance. In the meantime, scientists don't need to explain how a nonexistent consciousness arises any more than we need to explain how the Earth got to be flat or how the sun orbits the Earth. Proponents of this view are developing a growing following in science, but the idea is often difficult for people to accept.

The attention schema theory is a kind of illusionism. In the theory, the most perplexing property of consciousness—its ethereal, metaphysical nature—is not real. We think we have that property only because we are misinformed by an imperfect internal model.

In my experience, however, calling consciousness an illusion is the kiss of death for a theory. A handful of philosophers might know what you mean, while the rest of the world swats away your theory as ivory-tower foolishness: "How can consciousness be an illusion when I obviously have so much going on inside my head?"

The word *illusion* is so easily misunderstood that it can act as a barrier in discussions of consciousness. Here I'll describe three pitfalls in labeling consciousness as an illusion—but at the same time, I am not attacking the underlying concept of the illusionist approach, which I think is essentially correct.

I SAW A little boy and girl playing on the beach, digging in the sand. They must have been about 5 years old. The boy said, with great earnestness, "We shouldn't stay in the sun too long or we'll grow claws."

The girl was astonished. "Really?" she said, staring at him.

The little boy nodded solemnly. "It's true." He held up his hands and pantomimed pincher movements. "My mom told me."

This vignette is charming because the boy so obviously misunderstood a common metaphor. His mom must have told him that he'd turn into a lobster—meaning, he'd get a sunburn. His mind went to the claws instead of the boiled-red color.

Metaphors follow strict, implicit rules.[7] In a typical metaphor, the base (the lobster) is compared to the target (the sun-exposed human). Only one attribute is relevant. A lobster has many properties—claws, exoskeleton, compound eyes on stalks—but the person hearing the metaphor is expected to understand which single, key attribute is meant to be transferred. We all intuitively use metaphors in this way.

When a scientist says, "Consciousness is an illusion," I believe most people implicitly treat that statement as a metaphor. The base of the metaphor, a visual illusion, has many possible features. An object may look bigger than it really is or more tilted or farther away. A stationary object may look like it's moving. A convex surface may look concave. But when the word *illusion* is used in the context of a metaphor, it tends to mean one thing only. People isolate a key property: they equate "illusion" with "mirage." In a mirage, you think that something is present when it isn't. You don't just have the size or the details wrong—you have its very existence wrong.

For example, suppose a complaining friend tells you, "I swear, my manager's competence is an illusion." That person doesn't mean, "He's a competent manager, but in a slightly different way than you might expect. In fact, he might even be *more* competent than you expect." No, your friend means that his manager doesn't have any competence. In the context of a metaphor, to call something an illusion is to deny every aspect of its existence.

If you claim that consciousness is an illusion, most people don't take that to mean, "Consciousness is technically similar to a visual illusion because the information processing that we actually have in our heads is slightly different from what we claim to have on the basis of introspection, and we make that claim based on an imperfect internal model—all of which resembles a visual illusion." Instead, they take it to mean that nothing is present behind the illusion. There is no "there" there. Consciousness doesn't exist and nobody's home—a proposal that's inherently absurd to most people.

As I noted in earlier chapters, in the attention schema theory, consciousness is no mere mirage. It's a simplified, imperfect account of something real. The brain really does seize on information and process it deeply. When we claim to have a conscious experience, we're providing a slightly schematized version of that literal truth. There is, indeed, a "there" there. Consciousness resembles an illusion in a technical sense, which is why the attention schema theory is technically an illusionist theory. But consciousness is not an illusion in the way that most people understand the metaphor.

LET'S DIG DEEPER into illusions and how they differ from normal perception. Your visual system constructs internal models, simplified representations of the objects all around you. It does so automatically and constantly, as long as your eyes are open. That normal act of seeing isn't, by itself, an illusion. An illusion is a more specific case; it happens when the system makes a mistake. A small object looks big, or a straight object looks tilted. Something glitches. Most vision scientists understand the word *illusion* in this sense of a glitch or aberration from the normal.

The proposed attention schema is a normal internal model, constructed automatically and continuously. It gives you simplified information about your state of attention. I suppose it could malfunction, in which case that glitch in the internal model would constitute an illusion. Seeing consciousness in a puppet is an illusion. But I would not call a normally functioning internal model an illusion. Every internal model is a simplified version of reality, because reality has far more complexity and microscopic texture than the brain has any reason or ability to handle. If every simplified internal model qualifies as an illusion, then the terms *visual* and *illusion* are redundant, because all vision is an illusion. If that's the new definition of the word, then everything we see, hear, feel, and think is an illusion. Philosophers may have that all-inclusive definition in mind when they call consciousness an illusion—consciousness is part of the brain's imperfect understanding of reality. I see the logic of that argument, but I remain skeptical of the all-inclusive definition. The word loses its meaning if applied to everything.

MAYBE THE BIGGEST danger in calling consciousness an illusion is the risk that people will misinterpret the idea as a form of circular reasoning. To most people, an illusion is, by definition, a kind of conscious experience. If consciousness is an illusion, who's experiencing the illusion? The idea seems to use consciousness to explain consciousness.

This criticism can be frustrating, because it stems from a misunderstanding. To the illusionists, nothing in the brain experiences the illusion of consciousness. Instead, the brain claims to have consciousness on the basis of imperfect information.

I find it intriguing to talk to philosophers who say, as though startled at being asked, "Of *course* nothing in the brain has a subjective experience of the illusion of consciousness. That's not what I mean. Why, the concept would be circular otherwise!" Then I talk to anyone else, from any other walk of life, and they say, with the same startled look, "But of *course* an illusion implies that something conscious must be experiencing it! What else does that word mean? Why use the word if you mean something different? Why not just say that consciousness is a fruit bat and make up your own language?"

So I don't call the attention schema theory an illusionist theory, even though, according to the illusionists, it is. Our disagreement is about terminology rather than concept. Maybe consciousness is more like a caricature, because it distorts something real. I'm not sure I gain much clarity, however, by calling consciousness a caricature instead of an illusion. A sound byte or a slogan about consciousness is probably never going to be entirely intellectually satisfying.

An incredibly complicated, subtle machine has model-based knowledge about itself and its world. One particular internal model, the attention schema, gives us our intuitions about consciousness. Most of the properties we associate with consciousness actually and measurably exist in the brain in the form of the highest levels of cortical attention. Some properties that we associate with it, such as its metaphysical, ethereal nature, are the result of incomplete or schematic information in that internal model.

By virtue of those properties, is consciousness an illusion? You can call it that if you carefully define the word. I think my illusionist friends in philosophy have a smart approach and are precise in their definitions. I honestly do not mean to antagonize them. In fact, my hope is to show just how much our views can join together in one framework.

PHANTOM LIMBS

If you are ever unfortunate enough to lose a limb, you will probably experience a phantom, in which your missing limb feels as if it's still there. About 90 percent of amputees have a phantom sensation at least temporarily, and for some patients it can last for years.[8] The feelings of joint rotation, touch, pain, cold, and heat all remain. You can look and see that the limb is gone, but still feel every inch of it. Like the ghost of a limb, it extends invisibly from the body. Lord Nelson, the great British admiral who lost an arm in the battle of Santa Cruz de Tenerife, famously claimed that he had proved the existence of the afterlife because if his arm could have a ghost, then so could the rest of him.[9]

Phantom limbs are not just medical curiosities. The experience is extremely unpleasant. Imagine having a limb that seems psychologically real, and yet you can't scratch an itch, limber up the joints, or relieve pain. The phantom limb can be excruciating and debilitating, as well as bizarre and confusing.[10] For example, sometimes the phantom limb can "telescope," and the patient reports a phantom hand sticking out of the shoulder or a phantom leg that has become too short and doesn't touch the ground anymore.

The commonly accepted explanation for a phantom limb is that even though the limb itself is gone, the brain's internal model of it is still present.[11] A rich set of information, descriptive of a limb, lingers in the brain's circuitry. The phenomenon shows just how powerful an internal model can be. If it weren't for vision constantly proving the nonexistence of the limb, you would have no way to evaluate the truth. You would suppose that you actually did still have that limb.

The exact opposite of a phantom limb occurs in the clinical syndrome called somatoparaphrenia.[12] Patients who suffer stroke damage to parts of the parietal lobe might lose the internal model of a particular limb. They have the limb, they can see it, but it doesn't seem to belong to them. The neurologist Oliver Sacks, who had a special talent for describing clinical cases in humanizing ways, reported on a man who had lost the representation of his leg. In Sacks's account:

> "Easy" I said. "Be calm! Take it easy! I wouldn't punch that leg like that."
>
> "And why not!" he asked, irritably, belligerently.
>
> "Because it's your leg," I answered. "Don't you know your own leg?"[13]

One of the more powerful demonstrations of the internal model of the body comes from studying rubber hands. The first systematic demonstration of the rubber hand illusion was published in 1998 by Matthew Botvinick and Jonathan Cohen.[14] I experienced a later version of the experiment in the lab of Henrik Ehrsson in Stockholm. I sat in a chair in front of a desk and stuck my hand into a hole on the side of a shoebox. On top of the box sat a rubber hand, flesh colored, a little small for me, and not terribly realistic. I suppressed the urge to giggle.

The experimenter slipped a ring on my index finger, inside the box. The ring had a short, plastic rod sticking up, passing through a small hole in the roof of the box and attaching to a matching ring that was on the index finger of the rubber hand. Every time I lifted my own index finger inside the box, by mechanical coupling, the rubber hand on top of the box lifted up its index finger.

It only took me five or six finger lifts before the illusion hit me full force. Suddenly, that rubber hand became my hand. The feeling was shocking and profound. My body schema had incorporated a piece of rubber that I knew, cognitively, had nothing to do with me. I cannot even begin to describe the spookiness of that experience. The body schema is not visual information about a hand. It's not intellectual knowledge. It's not medical knowledge about which parts of the body connect to which. It's not a story I'm telling myself. It's an internal model, automatically computed, deep beneath cognition, beyond the reach of volition. It informs the cognitive brain, and we are captive to whatever it supplies. If it says that a rubber hand is my hand, then boom, I have an overwhelming gut certainty of it, even though, paradoxically, I know it's not true.

The body schema contains information that your brain needs to help control movement. It tells you what objects belong to your body. It depicts the shape of each body part, the hinged structure of the limbs, the size and extent, the way the arm moves and swings. What it doesn't contain is any information about the mechanistic, detailed structure inside the body. The arm schema says nothing about bone structure or tendon attachments, fast-twitch or slow-twitch muscle fibers, blood vessel paths, or the protein molecules inside muscle cells that cause contraction. If you close your eyes right now and talk about your arm—not your medical knowledge of how an arm is supposed to be, but your actual, in-the-moment insight into your arm—you can describe only superficial properties, not mechanistic details. That superficial description is supplied by your body schema.

FROM A PHILOSOPHICAL point of view, what is a phantom limb? I mean, what is the phantom itself? How should we classify it? It's not an object. It's not an energy field. Nothing is actually extending out of the stump. It *seems* like an invisible essence, the life force that's left behind after the physical flesh and bone is stripped away. It's similar to a ghost, as Lord Nelson pointed out. The culturally widespread belief in ghosts may derive from deep, internal models like the body schema, which feed that kind of construct to our higher cognition. And yet we can't just dismiss a phantom limb as a nonscientific superstition about ghosts. Something important is happening. Some useful process is present even before the limb is amputated and merely becomes unmasked after the limb is gone. The phantom is a simulation that we all have, and in the amputee, the simulation persists after the loss of the limb it is supposed to be representing.

I see a close analogy between a phantom limb and consciousness—between the body schema and the attention schema. One is the ghost in the body and the other is the ghost in the head. They are both simulations. They belong to different ends of the same object—a multipart model of the self. The body schema is a model of the physical self and how it works, while the attention schema is a model of another part of the self, the interacting neurons inside the skull and how *they* work. Both models leave out unnecessary mechanistic information. They are approximate, superficial, and yet indispensable. You could say that the attention schema is a specialized extension of the body schema.

I spent years studying the body schema, not just how the brain models the body itself, but also how it models a safety buffer of space around the body that warps and conforms as the limbs and head move, like a thick layer of invisible Jell-O.[15] The realization that the brain models the self in simplified, unrealistic, but profoundly useful ways led directly to the attention schema theory of consciousness.

The neuroscientist Olaf Blanke and the philosopher Thomas Metzinger make a particularly strong case for a connection between the body schema and consciousness.[16] In their account, bodily self-knowledge is a primordial, minimal form of consciousness, a knowledge of the self as an agent separate from the rest of the world. Obviously, this insight dovetails nicely with the attention schema theory.

THE GLOBAL WORKSPACE AND THE CONSCIOUSNESS KRAKEN

In my high school, we had a clique of popular kids. They were also gossip central. A lot of rumors and snippets circulated around the school at a low level, but if any tidbit of information reached the popular clique, it instantly became available to everyone, and suddenly the whole school knew about it. That, by the way, is also the global workspace theory of consciousness. Select information in the brain reaches a global workspace. It is broadcast everywhere and becomes available to guide our behavior and speech.

The global workspace theory was first proposed by Bernard Baars in the 1980s.[17] It has been elaborated by many others since then, especially by Stan Dehaene, to bring it into alignment with current knowledge about networks in the cerebral cortex.[18] In the theory, which I briefly described in Chapter 4, information percolates through a cortical hierarchy. Some of the information is able to rise in signal strength, outcompeting other signals and engaging the highest levels of processing, probably in the parietal-frontal networks. There, the information has entered the gossip central of the brain, the global workspace. It has achieved "fame in the brain," as Daniel Dennett calls it.[19] Information that has entered the global workspace has also entered consciousness.

The challenge of the global workspace theory is that it doesn't offer any specific explanation for why information, having arrived in the global workspace, gets the property of conscious experience attached to it. It's an incomplete theory—although, in a sense, that incompleteness is a feature, not a bug. One can collect data on the brain's anatomical substrate for consciousness without having to take sides in a philosophical debate. In some ways, it resembles the original scientific theory of consciousness, the theory of Hippocrates from 2,500 years ago, that the brain is responsible for the mind.[20] Hippocrates's theory was a milestone in science because it isolated the correct substrate, even if it didn't explain what consciousness is or how it comes about.

The attention schema theory offers a way to complete the picture. Let's suppose that the global workspace theory is correct, as far as it goes. Information can be boosted and selected by attention until it has a global impact around the brain. The attention schema theory posits that, in addition, the brain constructs a schematized model of attention. It builds its own naive metaphysical theory about what a global workspace is.

Where does that leave the relationship between the global workspace, the attention schema, and consciousness? I've tried to think of a good analogy to explain the complexities, and the best I can come up with is the kraken and the giant squid. The kraken is a mythological beast from Norse mythology, first mentioned in a thirteenth-century Icelandic saga, the Orvar-Odds saga.[21] The kraken is a huge squid with supernatural strength and ferocity, taken to destroying ships. Although the kraken doesn't exist, giant squid do. They live in the deep ocean, where they can reach lengths of 50 feet. They are rarely seen, are poorly understood, and never attack ships, since they die from the drop in water pressure when they get too near the surface. The kraken myth is almost certainly a distorted account of the giant squid.

Attention is analogous to the giant squid—a real if elusive phenomenon. Describing the highest level of attention in the brain as a global workspace is like describing the head of the giant squid as a global organ sac—a useful way for scientists to think about the anatomy. But neither of these explain consciousness—the kraken. To explain the kraken and its significance, it's not enough to dissect a squid and say, "There, that's a kraken." It isn't. The kraken is a supernatural, distorted version of the squid. It has a cultural and emotional impact that a mere squid could never have. A complete theory of the kraken should incorporate some understanding of the squid, but it must also include an understanding of the mythologizing process that leads to the kraken belief. Just so, a complete explanation of consciousness can't stop with attention and a global workspace. It must also incorporate that naive self model, the attention schema, which whispers to us about the consciousness kraken.

HIGHER-ORDER THOUGHT

The copy of my book that you're reading probably comes with a title on the cover, a title page on the inside, and some descriptive copy on the back. All of that extra stuff is not, strictly speaking, content. It's information *about* the content. It's meta-information that tags and labels. For the book to go into the world and have a useful effect, it needs that higher-order, meta-information attached to it.

The philosopher David Rosenthal suggested that something very similar happens to information in the brain.[22] When I look at an apple, for me to say, "I am conscious of the apple," it's not enough that my visual system processed information about the object. A higher-order thought must have been generated along the way and become bound to the apple information. This proposal is called the higher-order thought theory, and it fits into a general framework that is sometimes called "thinking about thinking," or metacognition.[23]

Scholars are still debating exactly what additional higher-order information might be added to the apple information to make us conscious of it. Maybe it's bookkeeping information.[24] A computer file contains primary information—the main contents of the file—and also higher-order, bookkeeping information displayed as an icon on the desktop. The icon stands for the file. It's a simple, compressed version of the file, minus the details. Maybe our visual system processes the apple and then compresses the information into a kind of icon, which becomes available for cognitive access and verbal report. In that perspective, we claim to be conscious of the apple because, to our higher cognition, the icon is somehow interpreted as an act of consciousness.

Another possibility for the higher-order information is a confidence rating.[25] Let's say you walk quickly past a fruit bowl with an apple in it. If you're not confident of what you saw, you would claim not to be conscious of the apple. If your confidence rating is high, you'll report that you're conscious of the apple. Maybe visual consciousness is visual information plus a high confidence that you actually possess that information.

A third possibility, emphasized by the philosopher Daniel Dennett, is that higher-order thought adds a more complicated, culturally learned layer of ideas.[26] We're all products of culture, and maybe one thoroughly learned cultural myth is that we have a soul inside of us that has subjective experience. If we were brought up in a radically different culture, we might not have that construct of consciousness, and then I wouldn't be writing this book. But because of an idea that may have been birthed tens of thousands of years ago in a cave or around a campfire and that subsequently went viral through the entire human population, we've all acquired the concept of consciousness, and our higher-order thinking attaches that concept to everything we do. The idea that consciousness is a complex of cultural memes has also been suggested by psychologist Susan Blackmore.[27]

The attention schema theory is a higher-order thought theory. It clearly belongs to the same category. And yet it has some differences from other examples of higher-order thought theories. In the attention schema theory, when the brain constructs consciousness of an apple, the visual information about the apple is linked to other, added information. But that added information does not comfortably fit the label of "higher order." In a sense, it's at the same level as the apple information. Both are representations of real items. Just as the brain constructs a representation of the shape of the apple, the color of the apple, and the spatial relationship between you and the apple, it also constructs a representation of your attentional focus on the apple. That attention schema is not intellectual, conceptual, or cognitive. In that sense, it isn't of a second order or higher order. It isn't culturally learned, and no upbringing could ever create it or unlearn it. We can't choose to turn it on and off. It evolved millions of years ago, long before our own species, before speech or human-level cognition. It's constructed beneath the level of language, though we do have some cognitive access to it and can talk about it. Like shape, color, visual movement, or spatial location, the attentional relationship between self and apple is another component of the larger dossier the brain compiles on that apple and its context in the world. The consciousness component is, in some ways, as basic as any of the other components. We believe we're conscious in the deepest, most intuitive, and most unreasoned way that we believe anything—because the brain constructs automatic models of its world and itself and has partial cognitive access to those models.

I agree, of course, that the brain *also* constructs other, higher-order thoughts. We do so for color, for example. We have emotional, cultural, and even political associations to the color red. Those extra connotations are layered on top of a more basic, automatic model of the color red, constructed deep in the visual system. In the same way, we have cultural and personal myths about consciousness—where it comes from, what function it serves, how it relates to philosophy and spirituality, and what happens to our consciousness when we die. But in the attention schema theory, underneath the cultural and conceptual associations that may vary from person to person, we also have an inborn attention schema that is more or less the same across all people and that gives us a common reference for consciousness.

ATTENTION AND AWARENESS

In 1890, William James, one of the pioneers of modern psychology, wrote: "Everyone knows what attention is. It is the taking possession by the mind, in clear and vivid form, of one out of what seem several simultaneously possible objects or trains of thought. Focalization, concentration, of consciousness are of its essence. It implies withdrawal from some things in order to deal effectively with others."[28]

James's quote, as famous and compelling as it is, shows a typical confusion between attention and consciousness. In his account, people contain something called "mind" or "consciousness," and attention refers to a concentration of it. In that perspective, attention is made out of consciousness. It's the central focus within a broader field of consciousness.

I think that James's view is how most people colloquially understand the word *attention*. But it's quite far from the modern, scientific use of the word and also far from how I use the word in this book. The gap between the colloquial and the scientific meaning of attention is probably the main source of confusion and disagreement around the attention schema theory. Because of that ambiguity, I never did much like using the word *attention*, but it has such deep roots in psychology and neuroscience that I have not yet found a better alternative.

In neuroscience, attention is a process in the brain whereby a representation (such as a visual representation of an apple) has its signals enhanced, competing representations have their signals reduced, and the enhanced signals have a correspondingly greater impact on systems around the brain. Attention is not just a focus on one central object; it can be spread and divided. If you think that you are conscious of something outside of your attention—that you are attending to A while also conscious of B, C, and D at the edges—that intuition is probably not correct; or at least, you are drawing on a colloquial definition of attention. By the scientific definition, you are probably attending to all of these items to some degree or switching attention rapidly among them. Colloquially, we may think of attention as a subset of consciousness, but scientifically, the relationship is entirely different. Attention is a layered set of mechanisms—a data-handling method—whereas consciousness is an inner experience that we claim to have. Attention is something the brain does; consciousness is something the brain says it has.

In 1890, James wouldn't have known anything about the technical side of information processing. Alan Turing would not spell out the principles of computing machines until the 1930s,[29] and Claude Shannon would not invent information theory until the 1940s.[30] It would never have occurred to James to think of attention as the action of a computing machine. He understood attention to be a state of consciousness, and the two remained entangled in the minds of scientists for the next hundred years.

The separation between attention and awareness was convincingly established for the first time in 1999 by a group of researchers in London, including Robert Kentridge, Charles Heywood, and Larry Weiskrantz, who studied a man with a remarkable disruption to his visual consciousness.[31]

Patient GY suffered a traffic accident when he was 8 years old and lost almost all of his primary visual cortex. After the accident, he was blind in the right side of space and on much of the left side as well. As an adult, when GY was seated in front of a screen and a dot was flashed at different locations, he could see it only when it appeared in a small part of the screen just to the left of center. He had no conscious vision outside that spared region. But when he was asked to point to the dot in his supposedly blind area, then seemingly miraculously, he could do it quite accurately. He couldn't consciously see it, but the visual information had gotten into his brain and could be used to guide his arm. He could also distinguish basic visual features, such as whether he was looking at a horizontal line or a vertical line, without ever experiencing visual consciousness. He said that he saw nothing, but simply *knew* what was in front of him with a high degree of confidence. The knowledge presented itself as cognitive certainty. This strange phenomenon is called blindsight, and it is a consistent result of damage to the primary visual cortex.

Kentridge and his colleagues studied patient GY to determine what he could and could not do in his "blind" field. They flashed a dot somewhere in the blind field. Then, immediately after, they showed a short line segment at the same location. GY's task was to tell whether the line segment was horizontal or vertical. His reaction time was short in that situation, because, apparently, the initial dot attracted his attention and primed his processing of the line segment. But when the experimenters flashed the dot in one place and then presented the line segment at a different place, his reaction time was slower. Apparently, the dot drew his attention to the wrong location. His attention then had to shift to the new location, causing a slight delay before he could answer.

This experiment was a watershed moment in the study of consciousness. It showed finally that the mechanisms of attention could exist even when the mechanisms of consciousness were broken. Clearly, attention is not just a local concentration of consciousness. It is a different property.

Over the past 25 years, a great many studies have confirmed that awareness can be separated from attention, not just in people with brain damage, but in healthy people as well.[32] People can pay at least minimal attention to dim or brief images, in the sense of focusing their processing resources on them and even reacting to them, while claiming that they see nothing at all.

In the absence of awareness, attention does not seem to work entirely normally, consistent with it losing a part of its control mechanism. To give an example, one of the most crucial skills that we use every day is the ability to *not* focus attention on something that we shouldn't. The world is full of objects vying for our attention, and sometimes we need to force our attention away from one object (such as a mosquito flying nearby) and keep our focus on another object (such as a book). In that circumstance, we may still pay some attention to the mosquito, monitoring it on the side, but we want to pay more attention to the book. This challenge is one of the more obvious examples of how an attention schema would be useful. To accomplish the task, the brain needs to know what state its attention is in at each moment, monitoring when too much attention is being pulled away from the book toward the mosquito. It also needs a good working model of the spatial and temporal dynamics of that pull in order to counteract it. Now suppose that the awareness system glitches. Some attention has been drawn to the mosquito, but the person is not subjectively aware of the mosquito. Attention and awareness have become dissociated. In that circumstance, if the attention schema theory is correct, then the brain doesn't know that its attention is leaking away to the wrong location. Its model of attention is incomplete. People in that circumstance should have trouble minimizing their attention on the mosquito. Counterintuitively, being unaware of the mosquito should cause more attention to be siphoned off to it and away from the book. Experiments from my lab and those of others, using distractor stimuli analogous to the mosquito and target stimuli analogous to the book, confirm this pattern of results.[33]

These many experiments may give the false impression that awareness and attention are easy to separate in the lab or that the separation happens routinely to people throughout a normal day. However, scraping awareness off of attention is like scraping paint off a wall. The two usually stick together. If you put the visual system at the edge of its capacity in controlled laboratory conditions, with brief or dim stimuli, then attention and awareness start to peel apart. But to find a visual stimulus so dim, so brief, so masked by other stimuli that people are unaware of it, and yet titrate the stimulus so that it is still strong enough to snag at least a little of the person's attention, is an almost impossible task. It took us several years of pilot experiments to drive a wedge between attention and awareness.[34] I have heard similar stories from other scientists.

In the attention schema theory, the whole point of awareness is to give the brain a running account of attention. Awareness therefore tracks attention closely, something like the body schema tracking the location of the arm. The two come apart only when the system is put under stress, struggling at the threshold of its ability.

INTEGRATED INFORMATION

When you pick up an apple, you can process the color, the shape, the smell, the smooth feel, the sound as you bite into it, the taste, your emotional response, and many other related aspects. Each of these features, in isolation, could be processed in your brain without engaging awareness. Even an emotional reaction can run subconsciously. But when consciousness is brought to bear, the components click together into a single, rich understanding.

This kind of observation has led to a consensus view—one of the very few—that consciousness is related to the massive integration of information throughout the brain.[35] Exactly how the two processes interact is debated. Some theories suggest that consciousness causes information to bind together. Others suggest the causality goes the other way: integrating information together into complex webs causes consciousness to arise. The best-known version is the aptly named integrated information theory of Giulio Tononi.[36] Proponents of his theory compute a number, phi, which represents how much integrated information is present in an object, be it a brain or a cell phone or anything else. As phi increases, so does consciousness.

Some of the details of these theories may not be compatible with the attention schema theory, but they all share a common insight: consciousness has a relationship to the integration of information. Here I want to focus on a fundamental point about integration that is easily overlooked. In a way, information itself can be sticky, and some kinds of information are stickier than others.

Imagine a scattering of dots on a piece of paper, most of them black, a few of them red. The red ones form a larger shape, the letter X. The shape stands out to your eyes. The shared information about color binds the dots together. The importance of this observation was first understood by the Gestalt psychologists of the early twentieth century, who studied the hidden rules that group visual images into larger wholes.[37] I don't mean that color information is literally sticky or that small atoms of information spontaneously bind into large information molecules. But in the particular information system of the brain, in a qualified, high-level way, color information can stick different pieces of shape information together. Color is a classic information connector.

Color is limited to the visual domain. It helps us group objects across the field of view, but obviously can't connect information across the other senses. Spatial location, however, is a more general connector. If you see the twitchy motion of a bird on a branch and hear a chirp that comes from the same location, you will tend to associate the two and treat them as a single object.[38] By linking visual information and sound information to the same piece of location information, like connecting tinker toys, your brain can join the whole set together into a larger, unified package.

One reason why location is so versatile as a connector is that it is a relational property. Location information doesn't describe a specific intrinsic feature of the bird, like color, shape, or loudness, which would limit its generality. Instead, it describes a relationship between you and the bird (it's 20 feet away and to your left), and that relationship can apply equally to all the specific features of the bird. Its color, shape, and sound all share *that* location. Location information is so useful for integrating other pieces of information that it has a special status in the brain as a facilitator. If spatial information were somehow suddenly removed from the brain, our perceptual world would fall apart into a jumble of separate features and senses.[39] Location information is like a fibrous binding material mixed into the brain's perceptual system, holding the pudding of other information together. One could justifiably formulate an "integrated information" theory of spatial location—not because integrated information generates location (which makes no sense), but because in the context of how the brain processes its world, location information is exceptionally sticky.

Even though information about location can link data across vision, audition, touch, and maybe even olfaction if you're a dog with a very good directional sense of smell, it's still a limited connector. Some information domains have no obvious spatial component. Emotions, thoughts, beliefs, imagination, mathematical insights—none of these are anchored to a specific spot in the space around your body. Location information cannot be used as a *universal* connector.

Can we find a kind of information in the brain that is so sticky that it acts as a universal connector, binding together any type of information with any other?

The attention schema—the information that describes your state of attention—could act as a universal connector. It's relevant to the bird you're looking at, the sound you're listening to, the thought you're considering, the emotion you're feeling. In the theory, the attention schema describes a general relationship between you and an object. In that sense, attention resembles spatial location, which is also a relationship between you and an object. But unlike location, the attention relationship is truly universally applicable. Anything concrete or abstract, perceptual or intellectual, can be the object of attention.

Imagine that the bird is still perched to your left. Let's analyze this one instant in time in which you're attending to the bird. Your brain constructs information about the bird's appearance and sound, your emotional reaction to the bird, perhaps even an intellectual idea about birds. You're attending to components A, B, C, and so on. Maybe you're attending mostly to the sound of its song and a little to its plumage, while even sparing a bit of attention to an unrelated itch on your arm. The next moment, your attention may be drawn elsewhere, but at this moment you are deploying your attention in a specific way across those specific components of your world. To represent the state of your world, yourself, and the relationship between them, your brain must construct an internal model of A, B, C, *and* of your attention and link them all together. By treating attention as a relational property of the world that is worth modeling, the brain constructs a central connector, the attention schema, to which all other information sets in the range of your attention will necessarily attach.

Accessing that interconnected web of information, your cognition learns, "There is a bird, it's colorful, it has a beautiful song, I'm happy, I must look up the species later, my arm itches—and all of these components are not separate, but are brought together under one umbrella because I have a subjective experience—a *consciousness*—that has taken hold of all of them at this moment."

The Gestalt psychologists dealt mostly in the sensory domain—studying color, shape, location, sound, and other sensory features—trying to understand how the perceptual world is bound together. The attention schema theory extends gestaltism by adding the ultimate connector. Consciousness pulls features together into a single, integrated whole—me, embedded in the world, at this moment in time. If location information is a binding compound, maybe it's like wood glue that works on some items and not others. Consciousness is more like a universal glue, binding across all information domains, working continuously no matter what is in the range of our attention. Without it, our whole world would break apart into a chaos of separate, floating components.

The attention schema theory, therefore, offers its own version of the integrated information theory. In that version, integrating information together, by itself, will not cause consciousness to arise in any direct way. Instead, information in the brain can be sticky, and some kinds of information are stickier than others. Color can help integrate information across the visual field. Spatial location can help integrate swaths of information across most of the sensory world. Presumably, many other kinds of information have varying degrees of stickiness. The stickiest information in the brain, the most universally connectable, is information about your attentional relationship to the items in your world. By definition, every item in the range of your attention shares that property. As a result of this "uber gestalt" property, consciousness presides over a massive integration of information in the brain.

CHAPTER 8

Conscious Machines

SIR ISAAC NEWTON, the great physicist and mathematician, tried his hand at alchemy.[1] Judging by the number of notebooks he filled, he seems to have put more effort into his failed attempt to make gold than his successful attempt to figure out gravity. One of his recipes called for Fiery Dragon, Doves of Diana, and Eagles of Mercury, all exotic names for real ingredients. I can imagine Newton and other alchemists pouring over their stained recipe parchments, mixing and heating, eagerly waiting for the outcome, only to be disappointed every time. They had no lack of theories, but their alchemical theories tended to be based on analogy or myth rather than logical inference. Practitioners could never say, "Given what is known about material properties, if you put A and B together, logically you will produce gold." At best, the aspiring alchemist would have to be content with the trial-and-error approach.

I don't mean to mock alchemy. That trial-and-error tinkering was arguably the origin of modern chemistry. And anyway, the great secret of alchemy was eventually cracked by modern science.

Here's the recipe, in case you'd like to try it yourself. Take an element heavier than gold, smash it in a particle accelerator, and eventually you'll produce a few atoms of gold. Or bombard a lighter element with nucleons, and eventually some of the components will stick together and form a gold nucleus. I don't know if anyone has tried it, but because the theory is physically sound, I know it will work. Elements much heavier and rarer than gold have been fabricated. The profit margin is not great, though—each atom of gold takes millions of dollars to produce—but it's the principle of the thing that counts.

Building a conscious machine reminds me of alchemy. A lot of the scholarship on consciousness is filled with vague analogy and sometimes even myth. If we put in enough complexity, or feedback, or widespread connections, or a resonating, thalamus-cortex loop, and stir it over a Bunsen burner, will the machine wake up and become conscious? Will it convincingly say, "I have a conscious experience of red and cold and myself"?

I think of the attention schema theory as an engineering answer to the alchemical mystery of consciousness. If you build a machine according to the theory, putting in the correct internal models and giving it cognitive and linguistic access to those models, then the machine will have the capabilities that you engineered into it. You won't have to hope that consciousness emerges out of an alchemical fusion of ingredients. The machine will think it has consciousness, claim to have consciousness, and talk about its consciousness, because you will have built the construct of consciousness into it.

The race to build conscious machines has had a slow start. There have been some preliminary attempts based on a variety of theories, but nothing verifiably resembling consciousness has emerged.[2] One damper on the enthusiasm may be that theories of consciousness tend to focus on the metaphysical feeling rather than on the practical benefits of consciousness. If you are a computer scientist interested in useful and marketable products, why waste your time on metaphysics? The attention schema theory, however, offers some practical benefits. In the theory, the brain evolved the construct of consciousness because it provided two substantial advantages: first, to improve internal regulation, and second, to serve as a foundation for social cognition.

Given a specific theory *and* specific practical benefits, artificial consciousness might now be poised to take off. We could potentially have it, in some form, in the next decade. Initial experiments in artificial consciousness are likely to be extremely limited. We might have machines that are conscious of visual images, but not yet conscious in any other way. Intelligent androids like Data on *Star Trek* or C3PO from *Star Wars*, which are not merely conscious but also omni-capable and conversationally brilliant, are much farther in the future. They are ultimately possible, but the technology is not yet here.

Before I wax poetic about a future of artificial, conscious companions, however, I need to start with a more fundamental question. If we build a machine that we think is conscious, how will we know if it really is? Can we test for the presence of consciousness?

IN 1950, THE mathematician Alan Turing proposed a way to test if a machine can think.[3] The test involves three people playing a game: a deceiver, a truth-teller, and a guesser. The three are seated in separate rooms and can communicate only through writing, to avoid any unintended personal clues passing between them.

At the outset, the guesser knows nothing about the other two people, except that one is a man and the other is a woman. The job of the guesser is to figure out which is which. The job of the truth-teller is to faithfully convey the correct answer to the guesser. The job of the deceiver is to confuse the guesser and convey the wrong answer. All three are locked in a complicated war of information. Since any question and any response is allowed, the conversation can dip into social innuendo and clever psychological manipulation. At the end of the game, the guesser makes a choice. If the guesser is right, the truth-teller wins. If the guesser is wrong, the deceiver wins. Without actually trying the game in reality, and I don't know of anyone who has, it's hard to know the typical success rate of a deceiver or a truth-teller. My guess is that the truth-teller has the advantage, because when you tell the truth you can't be tripped up in a contradiction. In any case, if the game were played many times over, the success rate of the different players could be estimated.

The next step is to replace the human deceiver with a machine deceiver. If the machine can win the game as often as a human, then the machine can think like a human. That's the original Turing test.

At first glance, Turing's test may seem overly complicated. It's certainly impractical, because it requires playing the game a great many times to accumulate statistics. But on closer look, it's also brilliant and ahead of its time. To perform well, the machine needs a theory of other people's minds. Even more than that, the machine needs to pass the false belief test. It needs to understand that other people, other minds, can contain false beliefs about the world, and it needs to keep track of who has what belief. It's a test of social cognition with a very high bar to pass.

Turing's test could be considered an early version of the Sally-Anne task, first proposed by psychologists in the 1980s.[4] I described the task in Chapter 5, but I'll briefly summarize it here. In the typical version, the psychologist testing your social cognition will tell you a story about how Sally is deceived by Anne. Sally puts her sandwich in basket A and leaves for a bathroom break, and while she's gone, Anne puts the sandwich in basket B. When Sally returns, which basket will she look in first to find her sandwich? If you can understand that Sally has a mind, that her mind holds beliefs, and that her beliefs can contradict reality, then you can solve the task. Sally will first look in basket A, where she originally put the sandwich. As trivial as the task seems to human adults with years of social experience, children under the age of about 5 can't solve it, and very few nonhuman animals show evidence that they can.[5]

The Sally-Anne task, with its mental models and tracking of false beliefs, is effectively a simplified, streamlined version of Turing's test. Turing may have thought he was writing about computing technology, but he was evidently a good social psychologist and 30 years ahead of his time. To test whether a machine can think, I would rather use Turing's complicated test than the stripped-down Sally-Anne test. It puts the machine through a more difficult challenge. To perform like a human, the machine would need spectacular language skills, a good working knowledge of everyday life, and an excellent ability to reason about other people's beliefs.

Turing never claimed that his guessing game was a test of consciousness. He never discussed its relationship to subjective experience, awareness, qualia, the inner feeling that comes along with information processing, or any other way to talk about consciousness. If a machine passes the original Turing test, we can be sure it's a spectacularly sophisticated computer. It must possess something like the normal *content* of human consciousness, or else it wouldn't hold its own in the conversation; but there is no guarantee that it has a conscious experience.

In the years since Turing's original publication, his test was appropriated and altered by a huge fan club in the field of artificial intelligence. The famous Turing test that most people have now heard of is actually quite different from the original, and also quite different from the Sally-Anne task. The modern version focuses on consciousness instead of social cognition. To tell if a machine is conscious, you should simply have a conversation with it. If you can't tell whether it's a machine or a person, then it passes. The new test is so much more streamlined than the original that it can easily be done in real life and has been tried many times. People have organized whole conferences around putting machines through this modernized Turing test, most famously the annual Loebner Prize competition, although, so far, no machine has gone on to convince scholars around the world of its consciousness.

If a machine could pass this modern Turing test, we still wouldn't know if it has an inner experience. A test of conversational skill is an illogical test of consciousness. Given how easy it is to convince some people that rivers and trees are conscious, it might not be all that difficult to trick at least a few people into thinking a machine is conscious, even if it's not. If we could build a machine that is truly conscious, it might still fail the Turing test. Consciousness does not always come with urbane, conversational sophistication. Heck, a 3-year-old child wouldn't pass the Turing test. Some adults wouldn't either. Neither would a pet dog, although most people are pretty sure that dogs have a conscious experience. It is just not a very discerning test. But the idea of the Turing test has become so familiar, so culturally engrained, that people seem to have accepted it as a practical solution. The common argument goes something like this:

I know I'm conscious because I have a direct experience of my own mind. But I'll never really know if other people are conscious. As much as I want to believe it, as much as I love my children and my wife and my cat, I can never directly experience their consciousness. It's impossible. I'll have to do the next best thing and provisionally *assume* that they are conscious because they act like they are. Similarly, I can't prove the existence of consciousness in a machine. That proposition is unprovable. But with the Turing test, I can do the next best thing. I can test whether the machine belongs to the same category as my wife, kids, and pets. I can find out if it acts so convincingly like a conscious being that I will have to provisionally assume it to be.

That's the standard argument, anyway. But I would like to deconstruct the mystique of the Turing test. From the perspective of the attention schema theory, we *can* know, with objective certainty, whether a machine has the same kind of consciousness that people have. And direct personal experience is *not* the only way—it's not even a very good way—to know about one's own consciousness.

"OF COURSE I'M conscious. I know I am, because I have a direct experience of it."

If that isn't the definition of a circular logic loop, I don't know what is. Consciousness *is* a direct experience. Therefore, the statement is tantamount to, "I know I'm conscious, because I'm conscious." As I've said before, the machine is captive to the information it contains. A particular internal model informs the machine that it has consciousness, and therefore it "knows" that it is conscious. The internal model informs the machine that its consciousness is without physical substance and forever private, and therefore it "knows" that its consciousness is unconfirmable by anyone else.

But an internal model is information, and information can be objectively measured. We don't need to rely on personal affirmation. In the attention schema theory, to go about determining whether a machine is conscious, we should probe its innards to find out whether it contains an attention schema, and we should read the information within the attention schema. We will then learn, with objective certainty, whether this is a machine that thinks it has a subjective conscious experience in the same way that we think that we do. If it has the requisite information in that internal model, then yes. If not, then no. All of this is, in principle, measurable and confirmable.

Measuring information in the human brain is difficult, but not physically impossible. Scientists already do it in a limited way. With a good set of electrodes on your head, scientists can read whether you plan to move your hand to the right or to the left.[6] With a high-quality MRI of the visual cortex, scientists can read whether you are looking at a face or a house.[7] Reading information of much greater complexity is not yet possible, but should be a matter of scaling up the technology. I'm sure that over time, the technology for reading information from the brain at truly high resolution will be developed (which is a disturbing thought). In principle, it should be possible to measure the information within the human attention schema, even if that kind of sophisticated information extraction from the brain lies in the distant future. My point is that in the attention schema theory, consciousness is not, by definition, always and forever private. It isn't true that "I know I'm conscious but can never really know if you are." Finding out whether a brain thinks it has consciousness in the same way that I do is a matter of developing the technology to read the information in it.

In the case of artificial intelligence, instead of a biological brain, looking under the hood and measuring consciousness should be a lot easier. It should be possible for engineers to tap into a human-made machine and measure its information content, given that we have much better tools for the job than we do for tapping into the brain and also given that somebody would presumably have the schematics for the machine.

The confusion arises when people ask, "I understand that you can measure information, but how will you know if the machine also has that inner feeling? Only the machine can directly experience its own consciousness, so who besides the machine can ever really know?" This question arises when your cognition consults your attention schema. That attention schema tells you that you possess a private, personal, immaterial feeling, to which only you can ever have access. But our hypothetical machine contains all the same components that you do, including an attention schema that contains the same information within it. On consulting its attention schema, it, too, would be informed that it has a private, personal, and immaterial feeling. Both you and the machine, constructed in the same way, are stuck in the same logic loop. You know only what you know and can report only the information within you—and that information can, in principle, be probed by someone else.

What I am suggesting here is that the seemingly unanswerable question—how can we ever know whether a machine is conscious?—can, in principle, be definitively answered once we have the right information-probing equipment. We shouldn't need to rely on the Turing test, which is at best indirect and at worst entirely irrelevant to the question of consciousness.

IN THE REST of this chapter, I'll consider the practicality of building a conscious machine using the attention schema theory as a guide. We need four components. First, the machine must have artificial attention—the ability to focus its resources and to shift that focus as it deeply processes one item and then the next. Second, the machine must have an attention schema—an internal model that describes attention in a general way and in so doing informs the machine about subjective consciousness. Third, the machine must have the right range of content. We could have a machine that is only visually conscious, for example—and I think that will probably be the first attempt—but with such a limited range, the machine would hardly resemble a human. Ideally, the machine would have much broader content. Fourth, the machine must have a sophisticated search engine that can access its internal models and talk about them, so that we can have a conversation with it and it can tell us about its consciousness. If we can build those four components, we will have a machine that possesses something like human consciousness.

Let's briefly look at these components one by one and assess how practical they are.

The first component is attention. The machine should have the ability to focus its processing resources and to shift that focus from object to object—from a nearby apple, to a doughnut, to a person standing at the far side of the room, to an entirely internal event such as a recalled memory.

Attention, in some form, already exists in artificial devices. A systematic management of resources is standard in any modern computer. In addition, many researchers have outlined or built computational models—computer simulations—that capture different aspects of humanlike attention.[8]

But for building a conscious machine, one has to get the right kind of attention, and that has not been convincingly tackled. Human attention has a specific suite of properties. When it is focused on an item, it enables you to process the item in detail and extract deeper meaning from it. Attention enables you to identify the item's affordances—can it be grasped, kicked, bitten? Attention empowers you to act—it enables you to make decisions about what to do next with the item. It allows you to store the item in memory for later use. The dynamics of attention mean that louder or stronger stimuli can snag the focus of your processing, but at the same time, attention is not at the mercy of the environment—it's also under your own control, so that an internal directive can shift attention from item to item. Above all, attention can shift easily between very different domains of information. It can be directed to spatial locations (the main property of attention that has been incorporated into artificial systems so far), but can also be focused on color, movement, taste, tactile pressure, or even something internal like a memory or an idea.

So far in the world of artificial intelligence, I don't know of any examples of attention at this level of complexity. One reason is that artificial intelligence tends to be domain specific. For example, consider the case of artificial face recognition. It's an important function with diverse applications, but it mainly sidesteps the issue of attention. A simple version of spatial attention can point the camera at the face that you want to identify, and then the recognition algorithm goes to work. The machine never needs to shift its focus of attention across multiple domains of information, from faces to sounds to internal thoughts, because it works only within the one domain of facial recognition. The system has essentially no relationship to attention except in the most trivial, overt sense of pointing the eyes at a target. Artificial attention is still technically possible, and many people are working on it, but so far it lacks the domain generality of human attention. This first component of a conscious machine is making genuine progress but needs more development.

THE SECOND COMPONENT the machine needs is an attention schema, the crucial internal model that describes attention in a general way and in so doing informs the machine about consciousness.

A colleague once told me that every desktop computer already has a kind of an attention schema, a set of information that monitors its distribution of processing resources. By my theory, then, every computer is already conscious. Another colleague told me that he could easily program up the whole theory in half a day. He could whomp up an artificial version of visual attention, attach it to a module that keeps track of its variables, and presto, that device should be conscious if I were right. A group of colleagues actually did build an artificial attention device with an attention schema to help it monitor and control its own attention.[9] The internal model did help make the machine more efficient at controlling its attention, just as expected. But is that machine really conscious?

The answer, I'm pretty sure, is no. None of these examples captures consciousness. An attention schema isn't a magic talisman that causes consciousness to emerge if you place it inside a computer. It's a packet of information. If the packet says something simple, such as "Variable A is within acceptable tolerance," and if your machine can verbalize that information, it wouldn't tell you, "I have a conscious experience!" It would tell you about variable A instead. There is nothing magical about the relationship between an attention schema and consciousness. If the attention schema tells the machine it has consciousness, then the machine will be informed that it has consciousness.

The human attention schema, having been shaped over hundreds of millions of years of evolution, has weird, biologically lumpy content. It depicts attention as an invisible property, a mind that can experience or take possession of items, a force that empowers me to act and to remember, something that in itself has no physical substance but still lurks privately inside me. The attention schema is more than a pointer to an object or a couple of lines of code. It builds a rich picture of attention and its predictable consequences. Build a machine with *that* kind of an attention schema, containing the weird, biologically messy lumps of the real one, and you'll have a machine that can claim to be conscious in the same ways that humans do. No major technological hurdle stands in the way of an artificial attention schema. It is, arguably, the easiest and certainly the most circumscribed of all the components to build into the machine.

A CONSCIOUS MACHINE would next need content. Building the rich, varied content of consciousness may be the most difficult part of the challenge, because the problem is open-ended. Ironically, the so-called hard problem (getting the machine to be conscious of anything at all) may be the easy part, and the easy problem (giving the machine a range of material to be conscious of) may be the hard part. My guess is that efforts toward artificial consciousness will begin with sensory input, especially vision, because so much is known about how sensory systems work in the brain and how they interact with attention. But I think we'd all be a little skeptical of a machine that is conscious of a black dot on a screen and nothing else. That might be a useful laboratory test case, but we would not consider that machine to be functionally conscious.

Even if we could give our machine a rich sensory consciousness, ultimately it should be able to go beyond the senses and incorporate abstract thought into its consciousness. Here the engineering problem becomes especially tricky. Little is known about how abstract thought intersects with the mechanisms of attention and awareness. We know that a person can pay attention to a thought, holding it in mind and focusing resources on it at the expense of other possible thoughts, even drawing attention away from sensory processing.[10] We've all had the experience of being so lost in thought as to lose track of what's going on around us in the external world. But neuroscience has not yet figured out what that thought-level attention entails mechanistically. Sorting out how to build a machine with those properties could take decades.

But I suspect that the hardest nut to crack will be emotion. It's the least understood domain of information in the brain. The best that can be said of it is that many brain structures responsible for emotion have been located.[11] People sometimes call them an emotional network or an emotional circuit, but the system isn't understood at the precision of a circuit.

One of the relevant brain areas is the hypothalamus, a little structure about the size of a walnut at the base of the brain. More than 100 years ago, the Swiss physiologist Walter Hess discovered that if he lowered an electrode into the hypothalamus and sent in a tiny whisper of electrical current to activate the neurons, he could induce what looked like emotional states.[12] His work has been confirmed many times since.[13] Fear, anger, lust, hunger—all of these states can be evoked by stimulating different spots in the hypothalamus.

Another central processor of emotion in the brain is the amygdala, a structure about the size of an almond (*amygdala* means "almond" in Greek). We have one on each side of the brain, and it revs up during emotional experience, especially when we associate a specific emotion with a situation or an image.[14] When you look at your least favorite politician and feel angry, that's your amygdala making the association between vision and emotion. Yet another brain area deeply involved in emotion is the bottom-most part of the prefrontal cortex, right behind the bony socket of the eye. It seems to play a role in making decisions based on emotional content.[15] While all of this neuroscience is fascinating, it tells us little about how to build artificial emotions or how to incorporate them into the mechanisms of attention and awareness.

One hint comes from an experience common to all people—emotions outside of consciousness.[16] We don't normally think of emotions as having a separate existence from consciousness, but it can happen. To give an everyday example, you might feel agitated, stressed, or angry at a low level and have no explicit idea of it. Then a friend points out, "Wow, what kind of mood are you in?" and you redirect your attention—and awareness—and realize, "Huh, I really *am* stressed." Some people are more aware of their emotions, some people less so, but everyone has moments of generating emotions without being aware that they are doing so. Any theory of emotional experience must take into account that distinction between emotional state and conscious experience.

The neuroscientist Joseph LeDoux, drawing on his pioneering work on the brain systems for emotion, formulated an elegant account of emotional consciousness.[17] Emotional states are organized outside of consciousness by structures deep in the brain, beneath the cortex—such as the hypothalamus and amygdala that I just mentioned. To become conscious of an emotional state requires information from those deeper structures to reach the cortical system, where it is integrated with cognitive information about consciousness. In this hypothesis, emotional consciousness contains two parts: information that defines the emotional state and information that defines consciousness. Just as the brain can compute the compound set of information, "I am conscious of the apple," so it can compute, "I am conscious of my emotion."

Even with this insight, the practical engineer is still stuck with the question, "What is the information that defines emotion?" The answer is that nobody really knows.

One hint comes from a famous theory of emotion proposed by the nineteenth-century psychologists William James and Carl Lange.[18] According to the James-Lange theory, emotion begins with a body sensation. The heart beats faster, the stomach secretes acid, the skin grows clammy. Then, detecting these bodily changes and assessing the context, the brain constructs a story: "I'm feeling anxious" or "I'm feeling excited." One of the best demonstrations of this effect was the famous Vancouver bridge experiment from the 1970s.[19] The experiment involved a woman stopping male pedestrians and asking if they would answer some survey questions. In some cases, the men were stopped in the middle of an anxiety-producing, swaying suspension bridge over a gorge. In other cases, the men were stopped on a solid, stable bridge. Afterward, the men were asked how attractive they found the female interviewer to be. Men who were on the anxiety-provoking bridge rated the woman as more attractive, whereas men on the safe bridge found her less so. Presumably, the suspension bridge caused the men's heart rate to rise and their skin to go a little clammy, changes that were misattributed to sexual attraction.

In modern psychology, the James-Lange theory is still considered partially correct. Part of emotion is anchored to the physical sensations in the body, and part of it involves a rich, high-level representation, the two parts interacting in a complicated and poorly understood way. To the extent that the James-Lange theory is true, it raises some interesting questions about machine emotion. Would an android need a stomach, sweat glands, and a heart to feel emotions the same way that people do? If the machine could never know the taste of food or the feel of digestion, never have adrenaline rushing through its body priming it to fight or flee, could it have real emotions? I suspect it can have processes very similar, though probably not identical, to human emotion. We might have to give it sensors throughout its body to ground its emotions in a physical substrate.

Early attempts at artificial consciousness will probably lack convincing emotion. These machines may be able to mimic an emotional tone of voice, but capturing the deeper truth of emotions, the way they are represented in the human brain, may be a long time in the future. A lot more basic science needs to be done. The Hollywood cliché of the emotionless android might turn out to be spot on, at least for a while.

THE FOURTH AND final component we need to build into our conscious machine is a talking search engine. We want a machine that can chat with us about its conscious experiences. Strictly speaking, talking isn't necessary for consciousness, but I think the goal for artificial consciousness, in the minds of most people, is a machine that has a humanlike ability to speak and understand. We want to have a good conversation with it.

It may seem like this particular aspect of the problem has already been solved, since we have digital assistants like Siri and Alexa. They can understand a verbal question, search a database, and then answer the question. Isn't that capability enough if we want a machine to talk about its internal states? But the problem is deceptively tricky. Siri operates mainly in the domain of language. You give it words, it searches for more words on the Internet, and then it gives you back words. If you ask for the nearest restaurant, Siri doesn't know what a restaurant is, other than as a statistical clustering of words. In contrast, the human brain can translate speech into nonverbal information and back again. If someone asks you, "How does the taste of a lemon compare with that of an orange?" you don't answer the same way an Internet search engine does. You don't rely on word association. You translate the speech into taste information and compare the two remembered tastes, then translate the answer back into words to give your answer. This easy back-and-forth conversion between speech and many other information domains is incredibly difficult to do artificially. As far as I know, the problem has not yet been solved in a systematic or general way. Google can, to some extent, translate information from visual images to words, but our conscious machine would need to correlate information across every imaginable domain.

GIVEN ALL THE promise and all the difficulties I've now mentioned, just how close are we to conscious machines?

The first attempts at visual consciousness—machines that incorporate both attention and an attention schema and bring them to bear on visual information—could be built over the next decade. I think it will take longer to give machines a range of content that resembles the vast range in human consciousness. To build a machine capable of seeing, hearing, tasting, touching, thinking abstract thoughts, and constructing emotions, capable of a single integrated focus of attention to coordinate within and between all those domains, *and* capable of talking about that full range of content is a long-term project. I will be surprised if we reach anything like it in the next 30 years (realistically, my estimate is more like 50 years). Still, I've been surprised before, and information technology is moving at a blinding pace.

People have dreamt of intelligent automatons for as long as mechanical technology has existed. Homer's *Iliad*, from almost 3,000 years ago, mentions self-wheeling tripods fashioned by the god Hephaestus. Five hundred years ago, Leonardo da Vinci engineered and built mechanical robots to entertain his patrons.[20] These days, everyone is familiar with Stanley Kubrick's HAL 9000 in the movie *2001: A Space Odyssey* and George Lucas's C3PO in *Star Wars*. A more recent movie, Alex Garland's *Ex Machina*, focuses entirely on how humans might interact with intelligent robots. Conscious machines have been on the human mind for so long that it's a little spooky to think how close we are to having them in real life. The saying comes to mind, "Be careful what you wish for."

WE DON'T YET have a good cultural model for how conscious machines might change us. Science fiction often makes accurate predictions about our future gadgets, but the social impact of those gadgets can be a lot harder to envision. For example, nobody predicted the social revolution that was caused by cell phones. At first we thought they would be an amazing convenience, like microwave ovens, but then they turned into the third hemisphere of the human brain. They restructured our political, economic, and social worlds. So if I can make one prediction with absolute confidence, it's that nobody can confidently predict the social impact of conscious machines.

Our science fiction tries to paint a picture of the future of artificial consciousness, but I doubt the picture will turn out to be very accurate. In the *Star Wars* franchise, conscious machines don't seem to add any ethically complex layer to society. They're simply second-class citizens. They form a staff of servile and technically capable slaves, charming but disposable, generally treated less respectfully than we treat our pets. Should conscious machines be treated so casually, or should they be given moral rights? The famous Isaac Asimov story "The Bicentennial Man" explores that question.[21] If we create conscious machines, do we have the right to control them and kill them? The film *Blade Runner*, based on a short story by Philip K. Dick, raises that moral question.[22] Novelists, directors, scientists, and philosophers have wrestled with these questions without resolution.[23] We may have to wait until we build truly humanlike artificial consciousness before we can develop a pragmatic ethic. The issues will probably look different than they do in science fiction, which is typically a disguised way to wrestle with today's dilemmas. Tomorrow's dilemmas may be unrecognizably different.

The biggest looming conundrum is probably not how well or how badly we will treat conscious robots, but how they will treat us. Conscious robots have become an apocalyptic cliché. In *The Terminator*, Skynet tries to kill us all. In *The Matrix*, machines enslave us. But again, that is science fiction, not reality.

Maybe because we've been overexposed to science fiction tropes, almost everyone misunderstands the concept of a conscious machine. Most people assume that turning on consciousness in a machine would be like turning a switch and waking it up. The machine would suddenly know about itself as an agent, a being separate from the rest of the world, and would be able to pursue a selfish agenda. It might start to kill off humanity as a competitor for resources. Machines with those capabilities would be a catastrophe for the human species. But here's the bad news. Superintelligent, autonomous, decision-making, agenda-pursuing, awake machines are already here in early form, and they are getting smarter all the time with an exponentially steep growth curve. They don't need consciousness to decide that people are obstacles and to murder us. A robotic truck with an autonomous agenda can run you over and kill you just as dead, whether or not it has a subjective experience of the squish under its tires.

The question to me, as a consciousness researcher, is this: What happens if we add that one extra piece—a self model that tells the machine it has subjective experience and that others can have it, too? How does it change the social equation?[24] I think it makes our future with intelligent machines more hopeful, not less.

Humans are consummately social animals. It defines us as a species. But a person can't be socially competent without a rich understanding of what consciousness is and an intuitive ability to attribute it to other people. That ability allows us to recognize other minds, understand each other's thoughts and feelings, and finesse the right way to respond. It's our human glue. It's the root of cooperation and society.

Imagine a world in which people have lost that deep, model-based knowledge about consciousness. Instead, we've been given alternative wiring. We don't become inert and unresponsive like people with clinical brain damage. We can still act as autonomous agents, learn, pursue agendas, and make intelligent decisions, but we do it all without any construct of consciousness. We become independent entities, each with a personal agenda, treating other people as objects. We can't cooperate meaningfully, even if we have the intelligence to know that cooperation is advantageous, because we can no longer understand that other people have minds, and therefore we can't guess at each other's mind states or properly coordinate our thoughts, goals, and actions. Murder becomes as casual as sweeping aside an obstacle, because we have no way of understanding the value of other minds. We become a world of intelligent, goal-directed monsters.

This is the world of unconscious but intelligent machines, and we're creating that world right now. I would rather live in a world of machines that know what consciousness is and that can attribute it to me and other people, just as I am grateful to live in a world where people can attribute consciousness to each other.

I'm not arguing that adding consciousness is a magic ingredient that will cause a machine to behave morally. We all know that the world contains some hateful people, including bullies and sociopaths. They're all conscious and probably know that the people they're hurting are also conscious. But they are misplaced tiles in a larger mosaic. They are always in the minority, because humans are fundamentally a prosocial species. Every time I walk into a crowded supermarket, I depend on the people around me being more or less aligned to the social matrix. Evolution has given us an imperfect, but statistically pretty good solution to the problem of social cohesion. What holds humanity together, what distinguishes us from a less cooperative and more antagonistic species, is our hypersocial thinking. That ability probably has many sources, but it would not even be possible without an internal model telling us that we are conscious and that others are, too.

Since the world is racing to create machines with humanlike or even superhuman intelligence and autonomy, I think we should also give them some humanizing traits so that they can have a chance at integrating with us. Why not try the same social, cooperative solution that evolution found for us? If the attention schema theory is correct, then building artificial consciousness is one of the most useful steps we can take to reduce the technological risks looming in our future.

CHAPTER 9

Uploading Minds

I DON'T WANT to live forever. I've always thought that if life has a purpose, it must be to contribute something useful to society, and living forever seems like it would tip the balance toward taking rather than giving. And it would get boring, too. Like playing a video game where you never die, all the jeopardy would be gone and a kind of listlessness would enter the soul. Well, that's just my opinion. I might change my mind as I get closer to the end. The truth is that many people, maybe even most, want to live indefinitely, and technology is moving toward that goal. I don't mean medical immortality, which does not appear to be physically plausible. I mean something called mind uploading—migrating the essence of a person, the contents of the brain, to an artificial platform. Sooner or later, people will invent an afterlife where human minds will live indefinitely in a simulated universe and interact with the real world by videoconference or robotic accessories. Life is full of strangeness, but of all the strange things I've encountered or considered, this looming computer-engineered afterlife has got to be the most bizarre.

If the theory I have worked on for the past 10 years is correct, then everything about the mind—memories, emotions, personality, even consciousness itself—is a product of physical mechanisms in the brain and can be copied. We should be able to scan the brain in enough detail to create a simulation of it, an artificial double of its information and algorithms, that could live on past the biological death of the person. In making that kind of statement, I'm not trying to belittle human complexity. As I've said before, the mind is a trillion-stranded sculpture made of information, constantly changing and beautifully complicated. But nothing in it is so mysterious that it can't in principle be copied to a different information-processing device, like a file copied from one computer to another.

Many commentators believe that mind-uploading technology is just around the corner.[1] I am not so optimistic about the time line. While it's true that anything to do with computing technology is advancing at an astonishing pace, anything to do with understanding or scanning the biology of the brain moves forward at a much slower, incremental one. However, I consider it inevitable that we will invent this technology. It may take centuries or it may come sooner than I expect, but technological trends and human motivation are pointed in that direction.

The challenge is fundamentally different from engineering artificial consciousness. It requires less knowledge about how the different components of a brain work together, because you don't have to build it from scratch. Instead, all you need to do is copy an existing brain. You don't need to know why it's wired up the way it is, as long as you can copy it faithfully. The challenge is that copying a brain requires scanning it at an almost unimaginable level of detail.

To BUILD A successful mind-uploading machine, we first need to figure out the minimum data set in the brain that captures the essence of the person. Most neuroscientists believe that computation in the brain is accomplished mainly by neurons connected to each other through synapses, the specialized contacts that allow information to flow in a gated way from one neuron to another. The human brain contains about

86 billion neurons.[2] It may contain about 100 trillion synapses, possibly even 10 times as many.

The so-called neuron doctrine, the principle that the brain works by means of neurons and synapses, was established a little more than a century ago by the Spanish scientist Santiago Ramón y Cajal.[3] He was one of the true geniuses of neuroscience, and for his work he won the 1906 Nobel Prize in physiology. By staining brain tissue and examining it under a microscope, he traced out the elaborate tendrils of neurons and developed the first real vision of how the brain works.[4] Information flows through neurons, passing along their delicate dendrites and terminals. That flow of information is gated by the synapses between neurons, blocked here or passed there, channeled into specific paths and circuits in the brain from input, through internal processing, to output. That overarching vision is essentially the same understanding that modern neuroscientists have. Cajal's beautiful drawings of individual brain cells are still shown in modern textbooks.

Inspired by the neuron doctrine, scientists and engineers have created artificial neurons and hooked them together into large networks to see how trainable or intelligent a mimic neural system can become.[5] The technology has revolutionized our world. Artificial neural networks have turned out to be extraordinarily powerful and adaptable. Internet search engines, digital assistants that seem to understand speech, self-driving cars, Wall Street trading algorithms, the guts of your smartphone—all of these now-standard parts of our world are powered partly by artificial neural networks.

Santiago Ramón y Cajal is not just the father of neuroscience, but also, in a way nobody could have anticipated, the founder of the current technology revolution.

The principle behind neural networks is that each neuron is extremely simple, but when they are connected together in large numbers, they are capable of great computational power. At a fundamental level, a neuron does nothing more than send a signal. Suppose neuron A is connected by a synapse to neuron B. When A becomes active, it fires off an electrical signal that travels down its length, reaches the synapse, hops across through a chemical messenger, and affects B. If the synapse is excitatory, then the signal boosts a simmering undercurrent of activity already present in neuron B, which becomes a little more likely to fire off its own signal. If the synapse is inhibitory, then the signal that hops across will quell neuron B, which becomes a little less likely to fire off a signal of its own. Synapses can also vary in strength, some of them able to pass a bigger signal, allowing neuron A to have a greater impact on neuron B, and some of them weaker such that neuron A might have almost no effect on neuron B. At the most reductionist level, this synaptic influence of neuron A on neuron B is all that happens, repeated trillions of times in a large interactive net. Each neuron receives inputs from as many as 100,000 other neurons. The job of each neuron is to count up those incoming signals and make a decision: given the barrage of input, all the excitatory and inhibitory chatter, the yeses and the nos pouring in at this moment in time, do the yeas or the nays have it? If the yeas have it, then the neuron fires off its own signal to affect the larger network. Each neuron's job is to make that one decision, over and over. Out of that seeming chaos of repeating simplicity comes complex computation.

Neural networks, whether biological or artificial, are exceptionally good at learning complex tasks. For example, if you want to train an artificial network to recognize faces, then you give it pixel input of faces from a digital camera, and it gives you back, out the other end, information about the identity of the face. In between is that maze of neurons and synapses through which the information flows. At first the network does a poor job, perhaps randomly associating faces with names. But every time it tries, it receives a training signal. As the network learns, it modifies which neurons are connected to which other ones, exactly how strong each connection is, and whether the connection is excitatory or inhibitory. In the end, the network learns the task by tuning up its pattern of synapses. If it sees a picture of Jim's face, whether shadowed or brightly lit, smiling or frowning, the machine associates that visual input with the correct output and can tell you, "That's Jim." Nobody knows ahead of time what the right pattern of synapses should be. You can't wire up every detail of a good face recognition device. Instead, the system learns by trial and error until a successful pattern of connections emerges to solve the task.

Given the past century of work on biological neurons and also the recent successes of artificial neural networks, most neuroscientists now believe that the essence of the brain lies in the pattern of connectivity among its neurons. In that view, if we could measure all the neurons in someone's brain, catalog which ones are connected to which, and characterize those synapses, we'd have the essence of the person. That hypothetical map of all the neurons and their synaptic connections is called a *connectome*,[6] a word intentionally analogous to the genome. The idea is that if scientists were able to map the human genome, an accomplishment once thought to be impossible, then they can tackle an even greater technical challenge and map the human connectome. Each person has his or her own unique connectome, defining a unique mind.

The last decade has seen some advances in mapping the connectome. The full connectome of a species of round worm, *Caenorhabditis elegans*, and more recently of the fruit fly have been published.[7] Scientists can also take a small piece of the cortex of a mouse, a few millimeters across, freeze it, slice it into extremely thin sections, scan each section, and reconstruct a high percentage of the neurons, their crisscrossing strands, and their synaptic connections.[8] The method has not captured every synapse yet, but a complete connectome of a few millimeters of a mouse's brain is, if not quite here, looking plausible for the future.

The National Institutes of Health (NIH) is currently sponsoring the Human Connectome Project, a massive worldwide effort whose ultimate goal is to map the connectome of the human brain. To study the human brain, MRI scanning methods can be used to trace networks of connections in ever-increasing detail.[9] This type of noninvasive scanning is a little more convenient to the participant, since it doesn't require freezing and slicing the subject's brain. A volunteer can lie in an MRI scanner for a few hours, and presto, a scan of amazing detail emerges. (I've had it done on my own brain many times. It's a little uncomfortable, fairly boring, and I tend to fall asleep, but the outcome is beautiful to see.) But these so-called human connectome maps are at a much coarser resolution than neurons and synapses. They capture larger-scale patterns, such as how one pea-sized part of the cortex connects to another. As the scanning technology improves, neuroscientists expect the connectivity in the human brain to be measured at a much finer resolution.

The progress I've described so far on artificial neural networks and on measuring the connectome in the real brain seems at first to paint an optimistic picture for mind uploading. It seems as though we know exactly what to measure in the brain, and we also know how to simulate neurons. Surely, given the rate of progress, we'll see brain uploading in the next decade or two? I do not share that optimism. It will happen, but not soon. In the next section, I'll explain why I think we are still very, very far from mind uploading.

LET'S TAKE A second look at the numbers I quoted earlier. The human brain contains about 86 billion neurons and perhaps around 100 trillion synapses, in a lowball estimate.[10] I don't know of any technology that can scan and measure 100 trillion of anything. The scale of the challenge is beyond today's gadgetry.

MRI machines can currently measure the brain down to a resolution of about half a millimeter, which is a truly amazing technological feat. But neurons are a lot smaller, and synapses are smaller yet. To detect a synapse, you would need to scan at a resolution of a micrometer—one thousandth of a millimeter. At that scale, you might begin to detect swellings on the neurons and guess that they might be synapses.

But even that fine scale would not be good enough. You would need to determine not just whether a hazy blob might be a synapse, but what kind of synapse it is and how strong a signal can pass through it from neuron to neuron. A physically larger synapse tends to pass a stronger signal, so as a first, crude method for measuring synaptic strength, you would need to measure synaptic size. You would need a clear, highly detailed image of each synapse at much better resolution than a micrometer. Your scanning device would also need to recognize whether the synapse is an excitatory or an inhibitory one. That information would probably require scanning for the presence of a specific kind of molecule within each synapse. No technology exists for that level of scanning. It may be possible through morphology—through the specific shape of a synapse—to guess more-or-less accurately what kind of synapse it is, rather than having to rely on a chemical analysis, but even that possible shortcut would require an incredibly fine resolution, maybe at a thousandth of a micrometer. We're not talking about an improvement in MRI technology. We're talking about an electron scanning microscope used on dead tissue or, for a living brain, new scanning technology that has not been invented yet.

The brain contains hundreds, possibly thousands of different kinds of synapses.[11] A gap junction, for example, involves a direct electrical connection between neurons. It is extremely fast and reliable and is crucial for the normal functioning of certain parts of the brain where precise timing is required. Another kind of synapse acts like a leaky spray bottle, sending out a puff of chemicals that affects a local patch of the brain rather than narrowly affecting one neighboring neuron. Some synapses contain more than one chemical transmitter, deploying each one under different circumstances. Some synapses are better at rapid change, a part of short-term learning, whereas others are more stable. All of these different kinds of connections between neurons, all these different shades and flavors, including many that we can assume have not yet been discovered, all their different dynamics, their speed and strength and adaptability, would need to be read out from the brain to construct its connectome.

Even if we could manage to scan all of that information from neurons and synapses, we'd still have to deal with the often-ignored glial cells.[12]

Most neuroscientists focus on studying neurons—hence the name of the profession. But the brain is packed with other cells that outnumber neurons 10 to 1. Glia were once dismissed as mere support cells, like a skeleton giving the brain its shape or like servant cells that supply neurons with their needs and clean up the by-products. But glia turn out to have properties directly related to the processing of information. Some glia secrete chemicals that influence neurons and synapses. Some even fire off the same kinds of electrochemical signals that neurons use to communicate. Their functions are poorly understood, but they are not as categorically different from neurons as once thought. The best that can be said at this time is that our knowledge of the processing in the brain has some humbling and gigantic holes.

All I'm saying is, don't be an early adopter of this technology when it comes out, because it will need a lot of refinement.

Suppose, in the future, a scanning technology is invented to register the synapses in your brain. A beta version might take into account only neurons, without the glia. It might have to simplify, categorizing synapses into only 100 major types. It might roughly digitize the strength of synapses—for example, allowing for 100 different possible increments of strength rather than capturing a more nuanced scale. The scanner might leave out the actions of hormones that spread diffusely through the brain. It might have an incredible success rate, correctly registering 99.99 percent of your actual synapses, and yet still miss enough of the detail in your brain to make a difference. The result might capture some echo of the original, but there is no telling what kind of ruined and terrible version of a mind might emerge from that process, once you build an artificial brain based on that data set. Would it be groggy, diseased, emotionally warped, and unable to concentrate?

It doesn't take much to disturb the normal balance of the brain. Just a trace of certain drugs can cause pain, confusion, hallucinations, and seizures. A concussion, which tears fibers and causes swelling, can lead to months or even years of foggy thinking and emotional instability. Even tiny irregularities can make a big difference. The simulated version of your brain had better work darn close to the original, or its experience will be hell. I'd wait until version 1000, when the kinks have been worked out. The guinea pigs on this one are going to be in for a rough ride.

When you look at the hurdles—the stupendous amounts of data that would need to be scanned from the brain, the submicrometer level of detail that the scanner would need to register, the basic neuroscience that is simply not yet known—it's tempting to give up on the whole project and assume that the dream of mind uploading is impossible. Yes, it is absolutely impossible, given today's technology. Not even a tune-up of today's technology would come within a light-year of mind uploading.

But at the same time, I am absolutely certain it will eventually get done. People have a way of solving technological problems. In 1916, Einstein predicted the existence of gravity waves.[13] He thought that the predicted effect was so absurdly small, about 10,000 times smaller than the nucleus of an atom, that it would be technically impossible to confirm. He could not imagine a machine of sufficient sensitivity, at any point, in any future. Almost exactly 100 years later, the machine was built and gravity waves were confirmed.[14] My guess is that new technology will be invented, new possibilities will open up, and mind uploading will become a reality. I can't predict when that might happen, because it depends on inspired people inventing unknown and currently unimaginable machines. If I had to put a number on it, given the rate at which fundamentally new scanning technology is invented, I'd guess at least 100 years if not substantially more, though I could be wrong if inspiration strikes someone sooner than that.

And I still don't recommend being an early adopter.

MIND UPLOADING HAS two components: first, scanning the relevant information from the brain you want to copy, as I have already explained;

and second, creating a working simulation of that brain. Let's pretend that the technical problems of scanning the brain have been solved. The crucial brain-scanning machine has been invented and all the necessary details of your brain have been captured. Now, you must use the data to construct a working simulation of your brain.

It may seem like this second part, the simulation, is the harder challenge, but actually, it is already effectively solved. The hardware is here. Artificial neurons and neural networks are standard. Adding extra kinds of synapses or subtler influences, such as hormonal effects, are not intrinsically difficult to simulate. Even networks containing millions of artificial neurons are straightforward. Companies around the world are currently trying to build systems that can rival the brain for complexity. The Blue Brain Project, for example, uses supercomputers to simulate massive brain-like collections of neurons. The Human Brain Project, the Allen Institute, Google Brain, DeepMind, Cogitai, and many other pioneering research groups are heading toward massive artificial neural network systems. Building a network with 86 billion neurons and 100 trillion synapses is still beyond today's technology. But the technology is scaling up rapidly, and especially with the advent of quantum computing, I have no doubt we will soon have sufficient firepower to simulate a network on the scale of a human brain.

This roaring progress in technology is part of what fuels the current optimism about mind uploading. A major piece of the puzzle, the most visible piece, is all but solved. But it is useful to keep in mind that an artificial neural network on the scale of a human brain, monumental though it may be, is not the same as an uploaded mind. The network by itself, without the right pattern of connections among those 86 billion neurons, is a useless pile of numbers. It's as if we're developing the printer that can print an artificial brain, and the material with which it can be printed, but we still need to tackle the problem of how to measure the right data from the real brain to feed into that printer. Otherwise we'll just print out meaningless mush.

To borrow a phrase from David Chalmers, I'd say we have a genuine hard problem of mind uploading: scanning a brain in sufficient detail.

LET'S MAKE A leap here and assume that we can simulate your brain. We've scanned it, and now we've re-created its neural networks. The next step is to embed that copied brain in a body. Without the body, it's not clear what kind of experience would accrue to your artificial brain as it is free-floating in a digital vat.[15] If that simulated brain actually has the same properties as a real person's, I think it would have a disorienting experience and maybe lose its sense of self. You derive your personal grounding from your body. You know where you are physically, where your arms and legs and torso are, and it gives you a primal anchor.[16] Without that anchor and without any contact to a world around you, without *embodiment*, I suspect that you would experience the kind of mental confusion reminiscent of a bad drug trip.

But now we need to decide what kind of artificial body to give you: a physical robot that can walk around the real world or a simulated body that inhabits a virtual world? A physical robot is limiting. It seems to me that implanting a simulated brain in a breakable, mortal body entirely misses the potential of mind uploading for flexibility and longevity.

Back in the days when my lab studied how the brain controls movement, we built a simulated human arm.[17] It wasn't a real, robotic arm. You couldn't actually shake its hand, and it couldn't pick up objects in the lab. It was a virtual arm, in the form of data in a computer. All we ever saw was a matrix of numbers on a screen. Based on scans of a real person's arm, it had everything it needed. Every bone, every tendon, every muscle. It had muscle forces, viscosity, inertia, and gravity. The muscles were made of separate muscle fibers, some of them fast-twitch and some slow-twitch. We gave the arm sensory neurons, alpha motor neurons, beta motor neurons, and gamma motor neurons. It would have cost us millions of dollars and years of development to try to build a robotic limb that mimicked a person's arm to the same level of detail, and we might never have succeeded. Instead, our virtual arm cost only a few thousand dollars for the computer and took a few months of effort.

If we could simulate a human arm in the mid 2000s, given the limited computing resources of the time, then it should already be possible to create a realistic, virtual human body, complete with the details of bones, muscles, nerves, and skin. I don't know of anyone who has done it—but it should be possible to create a super-realistic, video game avatar that resembles a human on the inside and out.

HAVING CREATED A simulated brain and a virtual body, our final task is to embed those elements in a virtual world. We can look to immersive video games for simulated, three-dimensional worlds, complete with their own physics. The technology is almost where it needs to be.

A truly convincing virtual world, with visual details that are fine-grained, sounds that travel through virtual space in realistic ways, wind that blows on virtual skin, perhaps even smells and tastes that are programmed to interact with virtual nasal passages and tongues—a virtual world at that level of detail does not yet exist. But it could. No new technology needs to be invented. It's a matter of scaling up the virtual environments that already exist. Processing power is the limitation here. If every supercomputer currently on Earth combined forces, I imagine there might be enough power to simulate a single human brain, its body, and a realistic three-room apartment for the artificial being to chillax. To accommodate many uploaded minds in a larger environment will require a major increase in computing power.

I realize that the picture I'm painting here for mind uploading is a strange, uneven mixture. Some of the technology is already here. Some of it is on the way, in the next 10 years at the most. Computers need more processing power—and they will soon have it. But some of the technology is far in the future, possibly centuries. The parts of the problem that are purely about information technology are much closer to realization, and the parts that relate to the biology of the brain are going to be slower to develop. But I see nothing to stop this mind-uploading technology from being achieved. Sooner or later, probably later, people's minds will be lifted from the biological brain and migrated to an artificial format.

I DON'T HAVE a dystopian view of mind uploading. It may have some major risks, but I think it also has great possibility. We humans, in our own messy way, tend to sort out what works and what doesn't, and I think mind uploading will be a cultural and ethical mess that sorts itself out eventually. Here I will briefly mention five potential pitfalls before describing what I think may be some benefits of mind uploading.

Potential Pitfall 1. In today's social climate, given how quickly we throw out old technology and adopt new gadgets every few years, mind uploading would be impractical. Your uploaded mind would be lucky to last 10 years before it was no longer compatible with the newest operating system. We'd be throwing out Grandma like we threw out WordStar, that ancient word processing program from the 1980s that is now so obsolete that few people even remember it existed. Our capitalist, consumer, high-turnover approach to information technology would have to change substantially before a mind-uploading platform would make any sense for extending anyone's life.

Potential Pitfall 2. The human brain has an extremely high memory capacity that has never been reached, although it is theoretically limited. As a simulated brain stored memory, its synapses would be reconfigured, and eventually it would run out of capacity to store new memories without disrupting older ones. I don't think anyone knows how large that capacity might be or when the workable limit might be reached. It could be on the order of centuries. Maybe, eventually, engineers will figure out how to add extra synapses to certain structures in the brain known to be involved in memory, such as the hippocampus, to give the simulated mind a memory booster every so often. Either that or a simulated mind will have to make do with a rolling memory window, in which events in the past few hundred years are clear and everything beforehand grows increasingly hazy.

Potential Pitfall 3. What would be the rights accorded to simulated and biological minds?[18] To get the technology to work, somebody's simulation is going to be put through existential torture, over and over, fine-tuning the method. If the test mind is close but not quite right, is it ethical to kill it and try again? If you make many simulated copies of the same person, does that mean each copy is less valuable and more expendable, or is each accorded separate moral rights? And do we care as much about the original, biological person, as long as some version of the person's mind is preserved? Or, to put it differently, what happens to the sanctity of life and of individuality when you've made three copies of yourself already?

Potential Pitfall 4. In many religions, the afterlife is held out as a reward for following the rules. That power of reward is often abused. The reward of heaven helped fuel the bloody and violent medieval crusades. Modern suicide bombers are enticed by promises of heavenly reward. But leaders who pedal a glorious afterlife are at a disadvantage: their product is unconfirmable. Now imagine the coercive power of an afterlife that is objectively verifiable. You can pick up your cell phone and text or talk to the people who are already there, find out how they like it, and even check out the Yelp reviews they posted. Whoever controls access to that confirmable afterlife controls the world. Almost every person on Earth would jump through hoops, even ethically dubious hoops—for many people, even starkly immoral ones—for a chance at immortality. I can easily imagine the emerging technology turning to dark political uses very quickly.

Potential Pitfall 5. Whose minds will be uploaded? Rich people? Brilliant people? Politicians with power? First come, first served? Resources could be severely limited, setting us up for an ethically complicated competition. Or it could be that memory and processing speed become so cheap in the distant future that it isn't an issue. Maybe the platform will wind up like YouTube, where anyone can join. Maybe it will have some version of net neutrality, or maybe some people will have faster bandwidth than others. I hope that whichever people are chosen to be uploaded are decent and inspiring, because with an extended life span they are likely to have an outsized effect on the rest of society.

EVEN SUPPOSING EVERYTHING goes well and the major pitfalls are avoided, it's still hard to understand what mind uploading would mean on a philosophical or cultural level. Consider one of the simplest questions: Is it really you?

You don't want to die, so one day you go into an uploading clinic, lie in a scanner for 5 hours while lights and sounds bleep around your head, and you come out a bit stiff in the joints but otherwise okay. Let's be generous and pretend that the technology works perfectly. It's been tested and debugged. It captures all your synapses in sufficient detail to re-create your unique mind. It gives that mind a standard-issue, virtual body that is reasonably comfortable, with your face and voice attached, in a virtual environment that is livable. Let's pretend all of this has come true.

Who is that second you?

The first you, let's call it the biological you, has paid a fortune for the procedure. And yet you walk out of the clinic just as mortal as when you walked in. You're still a biological being, and eventually you'll die. As you drive home, you think, "Well that was a bust and a waste of money."

At the same time, the simulated you wakes up in a virtual apartment and feels like the same old you. It has a continuity of experience. It remembers walking into the clinic, swiping a credit card, signing a waiver, and lying on the table. It feels as though it was anesthetized and then woke up again somewhere else. It has your memories, your personality, your thought patterns and emotional quirks. It sits up in a new bed and says, "Wow, it worked! I can't believe I'm here! *Definitely* worth the cost."

I won't call it an "it" anymore, because that mind is a version of you. We'll call it the simulated you. This "sim" you decides to explore. You step out of your apartment into the sunlight of a perfect day and find a virtual version of New York City. Sounds, smells, sights, people, the feel of the sidewalk underfoot, everything is present, with less garbage though, and the rats are entirely sanitary and put in for local color. You chat up strangers in a way you would never do in the real New York, where you'd be worried that someone might punch you in the teeth. Here, you can't get injured because your virtual body can't break. You stop at a café and sit at a picturesque wrought-iron table on the sidewalk, sipping a latte. It doesn't taste right. It doesn't feel like anything is going into your stomach. And nothing is, because it isn't real food and you don't have a stomach. You realize that you'll probably never have to go to the bathroom again. The visual detail on the table is imperfect. There's no grittiness to the rust. Your fingers don't have fingerprints—they're smooth, to save memory on fine detail. Breathing doesn't feel the same. If you hold your breath, you don't get dizzy, because there is no such thing as oxygen in this virtual world. You find yourself equipped with a complementary simulated smart phone, and you call the number that used to be yours—the phone you had with you just a few hours ago in your experience, when you walked into the clinic.

Now the biological you answers the phone.

"Yo," says the sim you. "It's me. I mean, it's you. What's up?"

"I'm depressed, you jerk. That's what's up. I'm in my apartment eating ice cream. I can't believe I spent all that money for bupkis."

"Bupkis? Dude! You would not believe what it's like in here! In some ways a little bland on the surface, but I'm sure I'll entertain myself. I passed a movie theater, and a bookstore, and we have money here so we can shop, thank goodness, and they say that the *Star Wars* simulator is so real that you're actually in the movie and you get a chance to be the Wookiee. And remember Kevin, the guy who died of cancer last week? He's here, too! He's fine, and he still has the same job. He Skypes with his old yoga studio three times a week to teach his fitness class. But his girlfriend in the real world left him for someone who's not dead yet, so he's a little bummed out. Still, lots of new people to date here."

I have to resist getting carried away by the delicious humor of the situation. Underneath the details lies a very real philosophical conundrum that actual people will eventually confront. What is the relationship between bio you and sim you?

I prefer a geometric way of thinking about the situation. Imagine that your life is like the rising stalk of the letter Y. You're born at the base, and as you grow up, your mind is shaped and changed along a trajectory. Then you let yourself be scanned, and from that moment on, the Y has branched. There are now two trajectories, each one equally and legitimately you. Let's say that the left-hand branch is the simulated you and the right-hand branch is the biological you. The part of you that lives indefinitely is represented by both the stem of the Y and the left-hand branch. The same way your childhood self lives on in your adult biological self, just so, the stem of the Y lives on in the simulated self. Once the scan is over, the two branches of the Y now proceed along different life paths, accumulating different experiences. The right-hand branch will die. Everything that happens to it *after* the branch point fails to achieve immortality—unless you choose to scan yourself again, in which case another branch appears, and the geometry becomes even more complicated.

What emerges is not a single you, but a topologically intricate version, a hyper-you with two or more branches. One of those branches is always going to be mortal, and the others have an indefinite life span, depending on how long the computer platform is maintained.

You might think that since the bio you lives in the real world and the sim you lives in a virtual world, the two will never meet and therefore should never encounter any complications from coexisting. But these days, who needs to meet in person? We interact mainly through electronic media anyway. The sim you and the bio you represent two fully functional, interactive, capable instances of you, competing within the same larger, interconnected social and economic universe. And you could easily find yourselves meeting over videoconference or other technology.

I suppose one solution to this existential confusion would be to store your brain scan data without turning it into an active simulation. Every so often, you could return to the clinic to make another backup copy of your brain. If you die in an accident or from disease or old age, then the most recent copy—or whichever favorite copy you specify in your will—can be activated and turned into a living simulation.

Another, rather dystopian way to trim back the number of simultaneous copies would be a law requiring that as soon as your mind is successfully uploaded for the first time and your simulation has been activated, then your biological body must be killed and recycled for crop fertilizer. Then we'd have rich and powerful moguls bribing the gatekeepers to break the law. They'd want 5 or 10 copies of themselves, plus the original. We'd need a special agent, like in a B-grade Hollywood movie, whose job is to track down and kill the illegal copies. It gets complicated quickly.

Then again, perhaps the system remains chaotic and freewheeling, with no restrictions on the number of versions of each person, causing a societal revolution in our concept of identity and individuality.

AT THE SIMPLEST LEVEL, mind uploading would preserve people in an indefinite afterlife. Families could have Christmas dinner with sim Grandma joining in on videoconference, the tablet screen propped up at the end of the table—presuming she has time for her bio family anymore, given the rich possibilities in the simulated playground. Maybe she's picked an athletic avatar for herself and is busy climbing simulated mountains.

It's this kind of idealized afterlife that people have in mind when they think about the benefits of mind uploading. It's a human-made heaven. But unlike a traditional heaven, it isn't a separate world. It's seamlessly connected to the real world. Think of how you interact with the world right now. If you live the typical modern lifestyle, then the smallest part of your life involves interacting with people in the physical space around you. Your connection to the larger world is almost entirely through digital means. The news comes to you on a screen or through earbuds. Distant locations are real to you mainly because you learn about them through electronic media. Politicians, celebrities, even some friends and family may exist to you mainly through data. People work in virtual offices where they know their colleagues only through video and text. Each of us might as well already be in a virtual world, with a steady flow of information passing in and out through CNN, Google, YouTube, Facebook, Twitter, and text. We live in a strange kind of multiverse, each one of us in a different virtual bubble, some of our bubbles merging and then separating in real space, but all of our bubbles connected through the global social network. If a virtual afterlife is created, the people in it, with the same personalities and needs that they had in real life, would have no reason to isolate themselves from the rest of us. Very little needs to change for them. Socially, politically, economically, the virtual and the real worlds would connect into one larger and always expanding civilization. The virtual world might as well be simply another city on Earth.

FOR TENS OF thousands of years, people used to preserve knowledge across generations by telling stories of the past. Ancestor worship was almost certainly an active part of that process.[19] By keeping alive the idea of our ancestors surrounding us and watching us, supporting us invisibly, people could hang onto some of their wisdom to help guide the living.

I doubt that when writing was invented in ancient Sumer,[20] around 6,000 years ago, anyone at the time understood how much it would transform our species. It was initially an economic tool, a way to keep track of who traded how much of what. But writing did something fundamental. It allowed people to speak directly across generations. Suddenly, the amount and the precision of data that could be accumulated over time exploded. Science and technology, in particular, depend on that accumulation of extremely precise information recorded by previous scholars. But the same could be said more or less of economic science, political theory, religious ideology, trends in art, and every other corner of our existence. Without writing, modern civilization would not have been possible, because it would not have accumulated in the same way, layer by layer, on top of the past.

Mind uploading would preserve more information at higher fidelity. It would also expand the kind of information that could be transferred—not just simple facts that can be written down in a book, but also subtleties of personality and skill that come through best when you talk directly to a person. In a spooky way, we would return to something like ancestor worship. But instead of a cloudy memory of the wisdom and exploits of the ancestors, rolled up in an oral tradition of storytelling, the actual ancestors would be here to talk to us directly. They wouldn't even have to whisper wise advice across the void. They could actively contribute to society through social media. In my view, this change to the way information transmits over time is likely to be the real source of revolution hidden in the new technology. Mind uploading would transform our species in a way that might surpass the evolution of speech and the invention of writing. The change might not always be constructive, however. Just as in the case of writing, television, or the Internet, when society takes on a new technology that increases the flow of information, that change inevitably comes with a rise in misinformation and harmful social memes.

To reduce the risk, we might decide to preserve the wisdom of our most accomplished people. An Einstein is different from an ordinary person because he has an exquisite set of synaptic weights—a lucky combination of what he was handed at birth and what he learned later. When you build and train artificial networks, something similar happens. You start with a large number of similar networks made out of the same number of neurons, train them, and see how they turn out. Some of them become geniuses that perform the desired task extremely well, and some are disappointments that get stuck in suboptimal states. The genius networks don't require any more bits of information to define them. After all, they're made out of the same number of neurons. Somehow they've settled onto a good pattern of synaptic connections. And yet you can't identify the good synapses. You can't say, "Ah, this particular inhibitory synapse has just the right weight. What beauty, what genius!" No, the engineer doesn't know specifically why the network functions so well. By training and by chance, the system has found an intangibly sweet set of synaptic weights. That's the network you want to save. At its most brutally pragmatic, mind uploading would preserve sweet networks. It would save people that have found a good set of connections toward a useful set of skills. Everyone learns intangible skills over a lifetime of experience. Imagine the power of preserving those skills.

Music would have evolved differently if Mozart had lived for another 200 years. Or Beethoven. Or, for that matter, Elvis or John Lennon. I'm not saying that the change would be good or bad, but I suspect that music would have changed less over time. History tells us that true revolution in music tends to happen with generational turnover, when the old guard fades and the new guard tries to break up the establishment. With successful mind uploading, some new voices might never have broken out, while old styles might have reached new heights.

Imagine how the English language would have evolved if Chaucer or Shakespeare remained with us, talking to us, writing, teaching, contributing. Most modern speakers of English find Shakespeare difficult and Chaucer unintelligible. Languages drift over time at a rate that has been much studied by linguists. As recently as 6,000 years ago, supposedly, a group of people spoke a lost language now called Proto-Indo-European, from which a vast collection of modern languages branched.[21] But linguistic drift would be reduced if people who spoke the language remained with us, interacting with the general population, mingling their voices with ours. The simulated minds might learn new ways of speaking—we all change our speech patterns over time—but linguistic drift would be slower if the older generations never died. With mind uploading, we might still be speaking a form of Proto-Indo-European 6,000 years later.

Not just language, but fashion, morality, entertainment, religion, and culture in general have a generational turnover. Despite our books, we reinvent the wheel on a regular basis. Imagine if culture were given greater inertia by people from the past who never died. The dynamics of cultural drift would be entirely different.

Imagine how politics would be affected. The philosopher and writer George Santayana said, "Those who cannot remember the past are condemned to repeat it."[22] And we do. We repeat history because we have generational amnesia. Political memory is not erased entirely; instead it fades with a half-life of around two or three generations. I don't mean that the intellectual knowledge goes away. Schoolchildren still learn the historical basics, but the immediacy, the emotional intensity of personal memory, is lost. History becomes theoretical over time. After the traumatic lessons of World War II, the whole world was attuned to the dangers of political populism and fascism. But the people who were actively engaged in those political events are now gone, the urgency of the memory has not been transmitted well across generations, and even the basic facts are denied by growing groups of people. Populism and fascism are creeping back in a way that would have been impossible 50 years ago.

I'm sure everyone can think of examples of the cyclic historical mistakes that civilizations make. No matter how well our books preserve bits of literal knowledge, every new generation that takes over from the previous one must reinvent the art and wisdom of how to live. What would happen if the old generation never died but remained on active duty? Ideally, political skill and wisdom would accumulate and we would avoid the constant doomed repeats. Maybe politics would become more like science in the sense of an ever-accumulating progress. On the other hand, for every Gandhi we'd love to live indefinitely, there'd be a Nero or a Hitler who would be a whole lot harder to kill.

Imagine how the university system would change. A faculty member does such a brilliant job for 30 years, teaching and serving on committees, that she's given the status of Emeritus. She keeps an office and continues to contribute. Then she dies. From Emeritus she is promoted to Mortem. She doesn't need a physical office anymore, just space on the upload platform. From there, she still teaches classes and joins committees via videoconference, contributing her unique knowledge and wisdom. If she's a history professor, she could be especially valuable as the centuries pass and new generations want to know how the world used to be from a firsthand account. Imagine the historical knowledge we would have available now if the state officials of ancient Egypt were still around and could tell us about their lives. And imagine how difficult it would be for young, incoming professors to earn tenure at a university clogged up with generations of dead scholars.

We don't normally think of the extreme elderly as active, contributing members of society. They're too frail, too few, and also too far behind on the latest trends. I know people—a diminishing crowd—who can't wrap their minds around a personal computer and don't know what the Internet is. Some of them won't carry a smartphone because the implications are too new to their brains. The rest of us are living in a boom of technological invention that pushes the older generation into obsolescence. We exist in an increasingly thin slice of the present, the overwhelming information of the moment. But whether good or bad, mind uploading would radically alter that pattern. The old folks wouldn't be a disappearing breed anymore. They'd be as present as anyone else and fond of their old familiar ways of living. Cultural innovation tends to come from the younger generation and be resisted by the older. An older generation that never fades would shift the demographics and almost certainly weigh the culture toward greater stability. Ironically, at the same time that a digital afterlife would drastically transform society, it might also make it more conservative and slow its rate of change.

Not only will older generations remain, but they are likely to accumulate power. There is no reason to think that the living will have any political, economic, or intellectual advantage over the simulated. Think of the jobs people have in our world. Many of them require physical action, and those are exactly the jobs that will likely be replaced by automatons. Taxi driver? Publicly shared, self-driving cars are almost here. Street cleaners? Checkout clerks? Construction workers? Pilots? All of these jobs are probably on the chopping block in the medium to long run. Robotics and artificial intelligence will take them over. The rest of our jobs, our contributions to the larger world, are done through the mind, and if the mind can be uploaded, it can keep doing the same job. A politician can work from cyberspace just as well as from real space. So can a teacher, a manager, a therapist, a journalist, a novelist, or the guy in the complaints department. The CEO of a company, a Steve Jobs type that has shaped up a sweet set of neural connections in his brain that makes him exceptional at his work, can manage without being physically present. If he must shake hands, he can take temporary possession of a humanoid robot, a kind of shared rent-a-bot, and spend a few hours in the real world, meeting and greeting. Even calling it the "real" world sounds prejudicial to me. Both worlds would be equally real. Maybe the better term is the "foundation" world and the "cloud" world.

The foundation world would be full of people who are mere youngsters—less than 80—and who are still accumulating valuable experience. Their unspoken responsibility would be to grow up, tune up, and gain wisdom and experience before joining the ranks of the cloud world. The balance of power and culture would shift rapidly to the cloud world. How could it not? That's where the knowledge, experience, and political connections will accumulate. In that scenario, the foundation world becomes a kind of larval stage for immature human minds, and the cloud world would be where life really begins.

LET'S TALK SPACE TRAVEL.

Star Trek gave us our first truly optimistic, culturally shared vision of the future: humanity will overcome war and disease and spread out over the galaxy as a peaceful, scientifically curious species in warp-drive starships. There we'll encounter and peacefully blow to smithereens various troublesome aliens. Whether you believe that vision of the future or a different one from the menu of science fiction options, we seem to share a cultural assumption that our future lies in space and that our species will colonize the galaxy.

But space travel is much more difficult than most people realize. The human body is not compatible with the toxic brew of outer space. Even if you give people a cabin with oxygenated air, a lethal rain of cosmic rays will pass right through almost every kind of shielding. The International Space Station, currently in orbit, is protected by the Earth's magnetic field, which deflects cosmic rays. Getting people to Mars, passing well beyond the protective influence of the Earth, may kill them. Getting them farther than Mars is exponentially more challenging.

If we could solve all the technical problems of maintaining people in space, there's still the problem of travel time. The nearest star system, Alpha Centauri, is more than four light-years away. Traveling at 10 percent the speed of light, far beyond any current space technology, would give us a 45-year trip, one way. If we could invent a magic spaceship that travels at 99.9 percent the speed of light, it would still take us more than 100,000 years to cross the galaxy. The human life span is too short for meaningful space exploration. We could have transgenerational space travel, and maybe we will. But in that case, the technical challenges of keeping people alive will only compound.

The dream of our species spreading out gloriously over the galaxy is almost certainly never going to happen. That future is a pipe dream. Our science fiction has gotten it wrong.

But there is a solution to space exploration. Uploaded minds don't need oxygen, atmospheric pressure, or organic food. They don't need to carry an Earth-like environment sealed in a cabin. They can also live indefinitely and can easily speed up or slow down their rate of conscious experience. Imagine a group of people—it could be hundreds, thousands, or even a duplicate of all the minds who have ever been uploaded. They exist on a simulation platform contained in a spaceship. Their responsibility is to explore. Because they live in a virtual world, they're not cramped inside the spaceship. They can spread out comfortably over a city or a simulated copy of Earth or any other simulated landscape. If they want to, they can create a virtual Starship *Enterprise* and live in it just like on TV, with big screens on the bridge showing their progress and little flashing buttons to amuse them. Their scientists and technicians would monitor the universe around them through viewports and instruments and guide their flight path accordingly. If they want to avoid a boring stretch of a few centuries between one star system and the next, they can simply turn a dial, slow down their processing speed, and make the experience seem as though only half an hour has passed—just enough time to comfortably adjust their trajectory on approach.

Having reached an alien star system, if they want to land on a planet and explore, they don't need a breathable atmosphere. All they need is a temperature that won't melt their equipment and gravity that won't crush them. If you are a member of that team, instead of putting on a space suit and leaving an actual crew capsule, you could transfer your neural network to a robotic lander, and off you would go around the new landscape, collecting samples and snapping pictures. When you are done, you would return to the simulated Earth or whatever the mother ship environment may be.

I suspect that most people today would prefer real people to colonize the universe. It doesn't seem like a proper space civilization if it's limited to virtual and robotic probes. We want someone like Neil Armstrong going out there and seeing a planet for himself, in the flesh. But I think that the only truly essential part of us, the part that defines a person, is the mind—and technology will one day allow human minds to travel freely through the universe.

I would argue for an inevitable, three-step progression. First, mind uploading will be invented. The psychological motivation for it, the desire to preserve the mind after the death of the body, is so great that scientists and engineers are already working toward that goal. Second, once mind uploading occurs, an obvious application is to send human minds, on artificial platforms, to places where human bodies cannot easily go. As a result, third, we will develop a full-blown spacefaring civilization that can travel great distances over immense timescales and disperse across the galaxy. The key to this future is the realization that the human mind is information, and it can in principle be migrated from a physical brain to an artificial system. The crucial moment in our development as a space-faring civilization is the moment we understand what consciousness is from an engineering perspective.

I WANT TO add one more thought about mind uploading, and it has to do with human sociality. People are drawn to each other. Human connectivity is in our nature. Yes, we have our share of loners, hermits, and agoraphobes, but most of us engage with a network of people. Writing letters wasn't enough for us, so we invented the telephone, then e-mail, and then texting. Now we walk around with smartphones glued to our hands, living off our social media feed. If we had the option of telepathy, a certain proportion of the population would seize on it and be in constant mind-to-mind contact with a network of a billion followers.

Outside of magical and pseudoscientific claims, humans have never directly shared thoughts with each other, mainly because we aren't born with a USB port in our heads. But a simulated mind would be made of information that is handled and manipulated on an artificial platform. One simulated mind could, technically, communicate directly with another.

I don't know how that technology would work. The beauty and simplicity of mind uploading is that engineers don't need to understand the details of how a brain works; all they need to do is copy it. If you copy it with good enough fidelity, it should work like the original. But to hook one mind directly to another would require a deeper understanding of how thoughts are computed. Which outputs do you use? Which inputs?

Any crude approach would merely skim random signals from one person's mind and inject them as meaningless noise into another. Given the limited state of knowledge in neuroscience right now, I can't even imagine how we would build that technology. Melding two uploaded minds together is an advance for the far-distant future, when much more is understood about the human brain. If it ever does happen, once again, we would see a fundamental change in the dynamics of information flow. Instead of a collection of simulated individuals, uploaded minds would turn into a central nexus of intelligence where human individuality is lost. People who are still living in the biological world would be mentally tiny, merely expendable fragments of brain that are busy accumulating experience, hoping for the chance to die and the privilege to meld into the collective.

Futures like this one usually sound horrifying to people—it sounds pretty horrifying to me, too—but I think the revulsion is mostly just a fear of the unfamiliar. The possible futures of mind uploading are so alien that we don't know how to wrap our emotions and thoughts around them.

I like to imagine a group of Cro-Magnon men and women sitting around a campfire at night some 30,000 years ago. One of them tells a story about an alien future. Their world will be leveled, the forests cut, the wilderness covered with artificial stones smothering the ground and stacked into giant square caves that block out the sky. The magnificent animals they hunt and paint on their walls will all be dead—the aurochs, the mammoths, the cave bears. The spiritual lifestyle of the hunter will be gone. Almost nobody will know how to make good spears or arrows or anything else with their own hands. Most people will sit inside, in dim light, pressing little clicking squares on a rectangular plate all day long, their bodies turning soft and slow. The air will be choked with human-made fumes, the outdoors roaring with a stupendous and continuous human-made din. I imagine the rest of the clan pulling faces and condemning this dystopian picture.

And yet here we are, and we don't mind it. Few of us would want to return to a pre-civilized, hunter-gatherer lifestyle. We know that our own world has its problems, but we don't want to change it too much. When we imagine a utopian future, we somehow always imagine a world more or less like our own, but with the worst problems solved and a few technological conveniences added. Any vision of the future that looks too different from the present must, by definition, be a dystopia. I think that people from all time periods will always have a similar view of the future.

But whether we approve or not, the world will change. Artificial intelligence and mind uploading will restructure it. People who live in that future will presumably muddle along just like we do in our world. They'll be used to it and probably wouldn't want to turn the clock back to our messy and limited existence.

When I try to look as far as I can into our future, the single most important change that I can see—the watershed moment in the history of our species—is the moment when people understand consciousness. Once we understand it from a pragmatic, engineering point of view, then a remarkable future becomes possible. In that future, mind is something precious, something to be nurtured, grown, and then saved, something that can be lifted from the original biological platform and migrated, duplicated, branched, maintained indefinitely, and even possibly merged with other minds. I see mind coming decoupled from biology and the line between artificial intelligence and human intelligence blurring beyond recognition. And I see that transcendent property of mind dispersing through space and exploring the galaxy, potentially for millions of years. The technology of mind may be our best path into a deep future.

APPENDIX

How to Build Visual Consciousness

Let's do a thought experiment. Our goal is to think through how to construct a machine that looks at an apple and is conscious of what it sees. This exercise has two potential uses. First, it serves as a tutorial on the attention schema theory. The underlying logic of the theory will be described in its simplest form. Second, I hope that the exercise will show engineers a general path forward for artificial consciousness.

I often use an apple in my examples because it combines the common visual features of a simple shape and a vivid color. We will give our machine a camera eye and a computer brain, and we want to know what to build into that artificial brain to make it insist that it has a subjective, conscious experience of the apple. The thought experiment is subject to a constraint: everything that we put into our hypothetical machine must be buildable. Some components may already be off-the-shelf technology. Some may not have been built yet, but if they are plausible given today's technology, then we're allowed to include them.

I'll present three different versions of our machine, each one a little more advanced than the previous, and by the final version I'll argue that the machine has essentially the same kind of consciousness that we have. I don't mean that the machine has a rich inner life. It would have no self-awareness, emotions, imagination, or goals. It would be conscious of one thing only—an apple. And yet just that small piece, by itself, would point the way toward how to build consciousness with a more inclusive content.

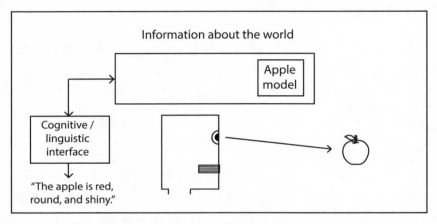

Figure A.1 A robot receives visual information through a camera eye and builds an internal model, or packet of information, about the apple. A search engine (the "cognitive/linguistic interface") allows the machine to answer questions based on its internal information.

Figure A.1 shows our robot looking at the apple. The box floating above the robot's head represents the information that is present in the device, and in this first diagram, the box contains only one item: visual information about the apple.

Visual processing is one of the most intensively studied topics in neuroscience. By now, the overarching outline of the human visual system has been traced, even if many mysteries and details have yet to be resolved. All we need to know for our build-a-brain project is that artificial visual systems have already been built, although they are simple and limited compared with the one in the human brain.[1] We will give our robot the capacity to take in a visual image through its camera eye and construct a rich, detailed set of information about the apple. The artificial visual system compiles information such as the color, shape, size, and location of the apple, constructing something like a dossier, a set of information that changes continuously as new signals come in. This packet of information is sometimes called an internal model. You could think of it as a simulation of the apple.

In the actual, biological brain, these simulations are not very accurate. It would be a waste of energy and a waste of processing resources for the brain to construct a detailed, scientifically accurate description of the apple. The color of the apple, for example, is partly a construct of the brain. In reality, the apple has a reflectance spectrum, not a color. The eye and the brain simplify that spectrum and assign a color. Color is a caricature, a quick-and-dirty proxy. In survival, efficiency is everything. You don't want computations so thorough that you can't respond to the world in a timely manner.

So far, nothing I've described violates the terms I set out at the beginning. If we had a camera and a computer, we could build a system like the one in Figure A.1. But is our robot *aware* of the apple?

In one definition of the word, yes. Consciousness researchers sometimes use the term *objective awareness* to mean that the information has gotten in and is being processed.[2] Yes, the machine in Figure A.1 is objectively aware of the apple. It contains the information.

But is it *subjectively* aware? Does it have a subjective experience of redness and roundness and shininess, the same way that you or I would when looking at an apple? Some scholars would still argue yes, consciousness is what it feels like to process information.[3] Since the machine is processing the apple, it is necessarily also subjectively aware of the apple. I call that type of theory the "information-adjacent" idea: consciousness is an inevitable by-product of processing information, just as heat is a by-product of electrical circuits. In that view, if you build a computing machine, consciousness will always emerge. If that is true, then we're done. We've successfully diagrammed a conscious machine, although consciousness remains unexplained.

But I don't think the machine is complete yet. To make the point, let's ask the robot. As long as we're doing a build-a-brain thought experiment, we might as well build in a linguistic interface, a kind of search engine like Siri (labeled in Figure A.1 as the "cognitive/linguistic interface"). It takes in questions, searches the database available within the machine, and answers. Because it is buildable with today's technology, we're allowed to include that search engine in our robot, even if this particular application of it is new.

We ask the machine, "What's there?"

Machine: "An apple."

Human: "What are the properties of the apple?"

Machine: "It's red, it's round, it's shiny, it's dented at the top, it's sitting at that location . . ."

It can answer these basic questions because it contains the relevant information. Moreover, its answers are rich and flexible because its knowledge of the apple is model based, and the model that it consults is a rich, if not perfectly accurate, description of the apple. We now have a machine that can look at an apple, process it, and make explicit statements about it. Everything about the machine is still fully buildable.

And yet, as a theory of consciousness, Figure A.1 is incomplete. To make the point, let's ask our machine an obvious follow-up question: "Are you aware of the apple?"

The search engine searches the internal model and finds no answer to this question. It finds plenty of information about the apple, but no information about awareness—about what awareness is or whether the machine has any of that strange property. The machine also lacks information about the self. After all, we asked, "Are *you* aware of the apple?" It lacks information on what this "you" quantity is. Our question is meaningless to the machine. At best it could answer, "Cannot compute." We might as well ask a digital camera, "Are you aware of the picture you just took?" Nothing that we've put into our machine so far would cause it to claim that it has consciousness.

I take it as a statement of fundamental logic that an information-processing machine cannot make a claim—it cannot output information—unless it contains the information that it is claiming. The machine in Figure A.1 contains no information about what consciousness is, let alone whether it is conscious of anything. We can't just hope that because we've put in some complexity, our machine will start claiming to be conscious. Given the components we've built into it, the machine contains no information about zebras, for example, and therefore it can't start talking about zebras. From an engineering perspective, hoping that our machine wakes up and claims "I am conscious" is a violation of logic. We will have to include more information in the machine to expand the range of answers it can give.

FIGURE A.2 SHOWS the next iteration of our build-a-brain project. Here I've added a second internal model: a model of the self. Again, the model is composed of information put together in the brain. The information might describe the physical shape and structure of the body, the so-called body schema.[4] It might include autobiographical memories, which

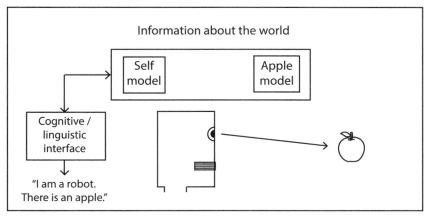

Figure A.2 The robot has been upgraded with a second internal model, a packet of information descriptive of itself.

play a major role in any psychological self model.[5] We know that the human brain contains something like a self model, probably very complex and probably spanning many brain areas. We're allowed to give our robot a self model, because giving it any arbitrary bundle of information is well within technical feasibility.

Now we ask our improved machine, "Who are you? Tell us about yourself."

With its new internal model, the machine has the requisite information to answer. It says, "I'm a metallic person; I'm six feet tall; I can bend my arms at the elbows and shoulders; I was born in a lab; I had a birthday five days ago . . ." The search engine can access and share the information available in its internal self model. Some of that information might be about the machine itself, and some might be about events in its past that help to define it.

Is the robot now conscious? Some might say it is. In many theories, consciousness is the same as self-knowledge. Some scholars who support this view focus on the body schema. In that approach, the most basic, primitive kind of consciousness is knowledge of the physical self and how it moves.[6] I know my body parts are mine: they are a part of me and separate in a fundamental way from other objects around me. I know my location in space: I exist here. I know my perspective: I see and hear and understand the world from my physical vantage point. Other scholars focus more on abstract, psychological self-knowledge.[7] In that perspective, I am conscious because I know my trajectory through life; I know my motivations and agendas. I can construct a running narrative to explain who I am and why I'm doing the things I'm doing.

The machine in Figure A.2 has plenty of self-knowledge. There is no technical barrier to building in all the self-knowledge we want, and I certainly agree that self-knowledge is important to consciousness. But again, I think that the diagram offers an incomplete account. To make the point, let's ask the machine, "What's the mental relationship between you and the apple?"

The machine is stuck. The search engine queries the internal models and finds plenty of information about the self, plenty of separate information about the apple, but no information about the mental relationship between them—no information about what a mental relationship is. It can't even parse the question. Equipped only with the components that we've given it so far, the robot can't answer the question. It cannot claim to be conscious of the apple, and consciousness is irrelevant to it.

THE ADVANTAGE OF internal models is that they monitor and make predictions about important items in the world. Our robot's world contains two obvious items: the apple and the robot itself. We have therefore given it two internal models: a model of the apple and a model of the self. But we've overlooked a third, subtle, hidden item in the robot's world: the computational relationship between the robot and the apple. To give the robot a complete description of its world, we need to build in a third internal model, a model of its own process of visual attention.

In a person, visual attention is necessary because a typical scene might include an apple, a plate, a table, a chair—sometimes hundreds of items—too many to process in depth. The brain must prioritize, focusing its resources on a limited set at a time. Only some items win the competition of the moment, their internal models shouting loud enough to have an impact on systems around the brain.[8] Unattended items go virtually unprocessed—we don't even register their presence—while attended objects are processed in depth. The brain can extract details and meaning from these selected items and decide how to react.

Sometimes when people use the word *attention* colloquially, they refer only to the one object that is at the center of interest. For example, a conversation might be the center of my interest, receiving most of my mental resources, while the apple I am eating is of peripheral importance. But when the word is used in the scientific sense, it is more inclusive. In that example, both the conversation and the apple are probably receiving attention. Their representations in the brain are boosted enough to be processed in a meaningful way. For the hundreds of objects or thoughts or sensory signals that are not attended, we do not even know they are present. When I talk about attention, I am referring to that larger relationship between the brain and the objects on which it is focusing resources.

Visual attention has been built artificially, although these artificial systems are much simpler than the biological, human version.[9] Because a visual attention machine is effectively off-the-shelf technology, already in existence, we are allowed to include the function in our build-a-brain project.

Figure A.3 shows the third and final iteration of the robot. Here, it's directing visual attention to the apple. Note that it also has three internal models, which provide it with a complete description of all three major components of its world. One model describes the apple, one describes the self, and one describes that third, invisible component of the scene, the attentional relationship between the self and the apple. That model of

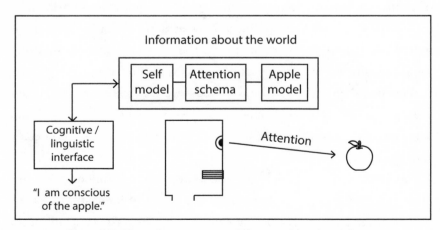

Figure A.3 The main components of the attention schema theory.
The robot contains a model of the apple, a model of itself, and a model
of the attentional relationship between the self and the apple.

attention is labeled the "attention schema," analogous to the body schema that describes the body.

We have a good idea of the information that should be contained in the model of the apple and in the model of the self. But what information should be contained in the model of attention? How would it usefully describe attention?

Its description can't be as simple as "The apple is the thing I'm attending to," or else it would be an empty model, a mere dotted line between two other models. An attention schema needs information about attention itself—about what it means to focus attention. But like the simplified internal model of the apple, a good attention schema would not waste resources on information about microscopic, technical details. In the case of the real brain, the technical details of attention include neurons, synapses, and competing electrochemical signals. In the case of our robot, the details might include wires, processing chips, and silicon logic gates. These are the nuts and bolts of attention that the system has no need to know about. Our machine needs a detail-poor, surface description of its attention to the apple. It might describe attention as a mental possession of something, a way of seizing on information and knowing it and understanding it in depth. It might describe attention as something that takes place roughly inside of oneself. It might describe some of the predictable consequences: attention confers the ability to react, to remember, to make decisions.

Now let's ask a question of the machine in Figure A.3: "What is the mental relationship between you and the apple?"

The search engine accesses the machine's internal models and replies based on the available information. It says, "I have a mental possession of the apple."

A promising answer, but we can probe deeper. "Tell us more about this mental possession. What are its physical attributes?" To make sure the machine can understand the question, we ask, "Do you know what physical attributes are?"

The machine has a body schema that describes its physical body, and it has an internal model of an apple that describes a physical object. So it can say, "Yes, I know what physical properties are. Solidity, weight, momentum, opacity—these are properties of objects." But based on the information available to it, it might say (if it has a good vocabulary), "My mental possession of the apple, the mental possession *in and of itself,* has no physically describable properties. Although it exists, it has no solidity, weight, momentum, or opacity. It doesn't exist on those scales. Instead, it's a nonphysical essence. In that sense, it's *metaphysical.* It does have a physical location, however—it exists roughly inside me. It's a part of me— the metaphysical part that allows me to have a mental grasp of something. I don't just process that the apple is red; I have a mental *possession* of the redness."

By adding an attention schema, we've constructed a machine that claims to have a consciousness suspiciously similar to the consciousness we humans claim to have.

Because we built the robot and know how it works, we remain skeptical. We push it harder by saying, "Thanks for the explanation, but you're still just a machine, and of *course* you're going to spew out all that information. You're merely accessing an internal model, your attention schema, and reporting the information within it. We're duly impressed that your model-based knowledge allows you to give such flexible, rich answers to questions. But still, you're merely reporting the information that you contain."

Hearing our complaint, the search engine accesses its internal models and finds nothing that matches that description. Remember, its internal models are too detail-poor to describe how the machine actually works. Given the limited information available to it, the machine replies, "I know nothing about internal models, information, or computations. Here is what I know: the apple is there, I am here, and I possess an awareness of the apple. Awareness itself is not a physical thing. It's also not a computation or a model, whatever those are. Instead it is a nonphysical property inside me that makes the apple vivid and present to me and permits me to react."

The machine not only claims to be conscious, but like most people, it rejects a mechanistic explanation of itself and believes in a metaphysical theory instead. It's built to do so, trapped by its own internal information. As long as it relies on introspection—and by that I mean, specifically, a cognitive search engine accessing the machine's internal models—it will always arrive at the same metaphysical self description.

Enough is known about machine vision that this limited system is probably buildable with today's technology, at least in a simple form. Going beyond vision will be more difficult until the technology advances further, but in principle, the same logic could apply to any item that the robot encounters. We could theoretically replace the apple with a sound, a touch, a memory, an emotional state, or the thought that 2 plus 2 equals 4. We could diagram the thought "I am aware of myself" or even "I am aware that I am aware." Consciousness of an apple is only the beginning.

NOTES

CHAPTER 1: THE ELEPHANT IN THE ROOM

1. My work on personal space and complex movement is summarized in two books. M. S. A. Graziano, *The Intelligent Movement Machine* (Oxford, UK: Oxford University Press, 2008); M. S. A. Graziano, *The Spaces between Us: A Story of Neuroscience, Evolution, and Human Nature* (New York: Oxford University Press, 2018).

2. The following references provide an overarching account of the theory. Other, more technical or data-oriented accounts are not listed here. M. S. A. Graziano and S. Kastner, "Human Consciousness and Its Relationship to Social Neuroscience: A Novel Hypothesis," *Cognitive Neuroscience* 2 (2011): 98–113; M. S. A. Graziano, *Consciousness and the Social Brain* (Oxford, UK: Oxford University Press, 2013); T. W. Webb and M. S. A. Graziano, "The Attention Schema Theory: A Mechanistic Account of Subjective Awareness," *Frontiers in Psychology* 6 (2015): article 500.

3. It is impossible to do justice to the vast new literature on the mechanistic, nondualistic approach to consciousness. Here I list a few examples and apologize to those many excellent authors whom I have left out. S. J. Blackmore, "Consciousness In Meme Machines," *Journal of Consciousness Studies* 10 (2003): 19–30; P. S. Churchland, *Touching a Nerve: Our Brains, Our Selves* (New York: W. W. Norton, 2013); F. Crick, *The Astonishing Hypothesis: The Scientific Search for the Soul* (New York: Scribner, 1995); S. Dehaene, *Consciousness and the Brain* (New York: Viking Press, 2014); D. Dennett, *Consciousness Explained* (Boston: Back Bay Books, 1991); K. Frankish, "Illusionism as a Theory of Consciousness," *Journal of Consciousness Studies* 23 (2016): 11–39; R. J. Gennaro, *Consciousness and Self Consciousness: A Defense of the Higher Order Thought Theory of Consciousness* (Philadelphia: John Benjamin's Publishing, 1996); O. Holland and R. Goodman, "Robots with Internal Models: A Route to Machine Consciousness?" *Journal of Consciousness Studies* 10 (2003): 77–109; T. Metzinger, *The Ego Tunnel: The Science of the Mind and the Myth of the Self* (New York: Basic Books, 2009).

4. D. Chalmers, "Facing Up to the Problem of Consciousness," *Journal of Consciousness Studies* 2 (1995): 200–219.

5. An earlier, insightful approach to consciousness that emphasizes internal models is: O. Holland and R. Goodman, "Robots with Internal Models: A Route to Machine Consciousness?" *Journal of Consciousness Studies* 10 (2003): 77–109.

6. G. Ryle, *The Concept of Mind* (Chicago: University of Chicago Press, 1949).

7. J. Joyce, *Ulysses* (Paris: Sylvia Beach, 1922).

8. D. Chalmers, *The Character of Consciousness* (New York: Oxford University Press, 2010); T. Nagel, "What Is It Like to Be a Bat?" *The Philosophical Review* 83 (1974): 435–50; J. R. Searle, "Consciousness," *Annual Review of Neuroscience* 23 (2000): 557–78.

9. R. A. Koene, "Scope and Resolution in Neural Prosthetics and Special Concerns for the Emulation of a Whole Brain," *Journal of Geoethical Nanotechnology* 1 (2006): 21–29; R. Kurzweil, *The Singularity Is Near: When Humans Transcend Biology* (New York: Penguin Books, 2006); H. Markram, E. Muller, S. Ramaswamy, M. W. Reimann, M. Abdellah, C. A. Sanchez, A. Ailamaki, et al., "Reconstruction and Simulation of Neocortical Microcircuitry," *Cell* 163 (2015): 456–92 ; A. Sandberg and N. Bostrom, "Whole Brain Emulation: A Roadmap," Technical Report #2008-3, Future of Humanity Institute, Oxford University, 2008.

CHAPTER 2: CRABS AND OCTOPUSES

1. Others have written compelling accounts of the possible evolution of consciousness, including linking consciousness to attention, though in different ways than I do. For example: C. Montemayor and H. H. Haladjian, *Consciousness, Attention, and Conscious Attention* (Cambridge, MA: MIT Press, 2015); R. Ornstein, *Evolution of Consciousness: The Origins of the Way We Think* (New York: Simon & Schuster, 1991).

2. O. Sakarya, K. A. Armstrong, M. Adamska, M. Adamski, I. F. Wang, B. Tidor, B. M. Degnan, T. H. Oakley, and K. S. Kosik, "A Post-Synaptic Scaffold at the Origin of the Animal Kingdom," *PLoS One* 2 (2007): e506.

3. Z. Yin, M. Zhu, E. H. Davidson, D. J. Bottjer, F. Zhao, and P. Tafforeau, "Sponge Grade Body Fossil with Cellular Resolution Dating 60 Myr before the Cambrian," *Proceedings of the National Academy of Sciences USA* 112 (2015): E1453–60.

4. D. H. Erwin, M. Laflamme, S. M. Tweedt, E. A. Sperling, D. Pisani, and K. J. Peterson, "The Cambrian Conundrum: Early Divergence and Later Ecological Success in the Early History of Animals," *Science* 334 (2011): 1091–7; A. C. Marques and A. G. Collins, "Cladistic Analysis of Medusozoa and Cnidarian Evolution," *Invertebrate Biology* 123 (2004): 23–42.

5. H. R. Bode, S. Heimfeld, O. Koizumi, C. L. Littlefield, and M. S. Yaross, "Maintenance and Regeneration of the Nerve Net in Hydra," *American Zoology* 28 (1988): 1053–63.

6. R. B. Barlow Jr. and A. J. Fraioli, "Inhibition in the Limulus Lateral Eye in Situ," *Journal of General Physiology* 71 (1978): 699–720.

7. K. Hadeler, "On the Theory of Lateral Inhibition," *Kybernetik* 14 (1974): 161–5.

8. S. Koenemann and R. Jenner, *Crustacea and Arthropod Relationships* (Boca Raton: CRC Press, 2005).

9. B. Schoenemann, H. Pärnaste, and E. N. K. Clarkson, "Structure and Function of a Compound Eye, More Than Half a Billion Years Old," *Proceedings of the National Academy of Sciences USA* 114 (2017): 13489–94.

10. R. Gillette and J. W. Brown, "The Sea Slug, *Pleurobranchaea californica*: A Signpost Species in the Evolution of Complex Nervous Systems and Behavior," *Integrative and Comparative Biology* 55 (2015): 1058–69.

11. C. R. Smarandache-Wellmann, "Arthropod Neurons and Nervous System," *Current Biology* 26 (2016): R960–R965.

12. S. Koenig, R. Wolf, and M. Heisenberg, "Visual Attention in Flies—Dopamine in the Mushroom Bodies Mediates the After-Effect of Cueing," *PLoS One* 11 (2016): e0161412; B. van Swinderen, "Attention in *Drosophila*," *International Review of Neurobiology* 99 (2011): 51–85.

13. D. H. Erwin, M. Laflamme, S. M. Tweedt, E. A. Sperling, D. Pisani, and K. J. Peterson, "The Cambrian Conundrum: Early Divergence and Later Ecological Success in the Early History of Animals," *Science* 334 (211): 1091–97; B. Runnegar and J. Pojeta Jr., "Molluscan Phylogeny: The Paleontological Viewpoint," *Science* 186 (1974): 311–17.

14. J. Kluessendorf and P. Doyle, "*Pohlsepia mazonensis*, an Early 'Octopus' from the Carboniferous of Illinois, USA," *Palaeontology* 43 (2000): 919–26; A. R. Tanner, D. Fuchs, I. E. Winkelmann, M. T. Gilbert, M. S. Pankey, A. M. Ribeiro, K. M. Kocot, K. M. Halanych, T. H. Oakley, R. R. da Fonseca, D. Pisani, and J. Vinther, "Molecular Clocks Indicate Turnover and Diversification of Modern Coleoid Cephalopods during the Mesozoic Marine Revolution," *Proceedings of Royal Society, B, Biological Sciences* 284 (2017): 20162818.

15. P. Godfrey-Smith, *Other Minds: The Octopus, the Sea, and the Deep Origins of Consciousness* (New York: Farrar, Straus and Giroux, 2016); S. Montgomery, *The Soul of an Octopus* (New York: Atria Books, 2015).

16. A.-S. Darmaillacq, L. Dickel, and J. A. Mather, *Cephalopod Cognition* (Cambridge, UK: Cambridge University Press, 2014); D. B. Edelman, B. J. Baars, and A. K. Seth, "Identifying Hallmarks of Consciousness in Non-Mammalian Species," *Consciousness and Cognition* 14 (2015): 169–87; J. N. Richter, B. Hochner, and M. J. Kuba, "Pull or Push? Octopuses Solve a Puzzle Problem," *PLoS One* 11 (2016): e0152048.

17. B. Hochner, "An Embodied View of Octopus Neurobiology," *Current Biology* 22 (2012): R887–92.

18. P. M. Merikle, D. Smilek, and J. D. Eastwood, "Perception without Awareness: Perspectives from Cognitive Psychology," *Cognition* 79 (2001): 115–34; R. Szczepanowski and L. Pessoa, "Fear Perception: Can Objective and Subjective Awareness Measures Be Dissociated?" *Journal of Vision* 10 (2007): 1–17.

CHAPTER 3: THE CENTRAL INTELLIGENCE OF A FROG

1. E. Knudsen and J. S. Schwartz, "The Optic Tectum, a Structure Evolved for Stimulus Selection," in *Evolution of Nervous Systems*, ed. J. Kaas (San Diego: Academic Press, 2017), 387–408; C. Maximino, "Evolutionary Changes in the Complexity of the Tectum of Nontetrapods: A Cladistic Approach," *PLoS One* 3 (2008): e3582.

2. D. Ingle, "Visuomotor Functions of the Frog Optic Tectum," *Brain, Behavior, and Evolution* 3 (1970): 57–71.

3. R. W. Sperry, "Effect of 180 Degree Rotation of the Retinal Field on Visuomotor Coordination," *Journal of Experimental Zoology Part A: Ecological and Integrative Physiology* 92 (1943): 263–79.

4. C. Comer and P. Grobstein, "Organization of Sensory Inputs to the Midbrain of the Frog, *Rana pipiens*," *Journal of Comparative Physiology* 142 (1981): 161–68.

5. B. E. Stein and M. A. Meredith, *The Merging of the Senses* (Cambridge, MA: MIT Press, 1993).

6. C. Comer and P. Grobstein, "Organization of Sensory Inputs to the Midbrain of the Frog, *Rana pipiens*," *Journal of Comparative Physiology* 142 (1981): 161–68; D. Ingle, "Visuomotor Functions of the Frog Optic Tectum," *Brain, Behavior, and Evolution* 3 (1970): 57–71.

7. B. E. Stein and M. A. Meredith, *The Merging of the Senses* (Cambridge, MA: MIT Press, 1993).

8. T. Finkenstadt and J.-P. Ewert, "Visual Pattern Discrimination through Interactions of Neural Networks: A Combined Electrical Brain Stimulation, Brain Lesion, and Extracellular Recording Study in *Salamandra salamandra*," *Journal of Comparative Physiology* 153 (1983): 99–110.

9. B. E. Stein and N. S. Gaither, "Sensory Representation in Reptilian Optic Tectum: Some Comparisons with Mammals," *Journal of Comparative Neurology* 202 (1981): 69–87.

10. H. Vanegas and H. Ito, "Morphological Aspects of the Teleostean Visual System: A Review," *Brain Research* 287 (1983): 117–37.

11. P. H. Hartline, L. Kass, and M. S. Loop, "Merging of Modalities in the Optic Tectum: Infrared and Visual Integration in Rattlesnakes," *Science* 199 (1978): 1225–29.

12. S. P. Mysore and E. I. Knudsen, "The Role of a Midbrain Network in Competitive Stimulus Selection," *Current Opinion in Neurobiology* 21 (2011): 653–60.

13. R. H. Wurtz and J. E. Albano, "Visual-Motor Function of the Primate Superior Colliculus," *Annual Review of Neuroscience* 3 (1980): 189–226.

14. M. I. Posner, "Orienting of Attention," *Quarterly Journal of Experimental Psychology* 32 (1980): 3–25.

15. E. F. Camacho and C. Bordons Alba, *Model Predictive Control* (New York: Springer, 2004); R. C. Conant and W. R. Ashby, "Every Good Regulator of a System Must Be a Model of That System," *International Journal of Systems Science* 1 (1970): 89–97; B. A. Francis and W. M. Wonham, "The Internal Model Principle of Control Theory," *Automatica* 12 (1976): 457–65.

16. M. S. A. Graziano and M. M. Botvinick, "How the Brain Represents the Body: Insights from Neurophysiology and Psychology," in *Common Mechanisms in Perception and Action: Attention and Performance XIX*, ed. W. Prinz and B. Hommel (Oxford, UK: Oxford University Press, 2002), 136–57; N. Holmes and C. Spence, "The Body Schema and the Multisensory Representation(s) of Personal Space," *Cognitive Processing* 5 (2004): 94–105; F. de Vignemont, *Mind the Body: An Exploration of Bodily Self-Awareness* (Oxford, UK: Oxford University Press, 2018).

17. H. Head and G. Holmes, "Sensory Disturbances from Cerebral Lesions," *Brain* 34 (1911): 102–254; G. Vallar and R. Ronchi, "Somatoparaphrenia: A Body Delusion. A Review of the Neuropsychological Literature," *Experimental Brain Research* 192 (2009): 533–51.

18. A. M. Haith and J. W. Krakauer, "Model-Based and Model-Free Mechanisms of Human Motor Learning," in *Progress in Motor Control: Neural Computational and Dynamic Approaches, Volume 782*, ed. M. Richardson, M. Riley, and K. Shockley (New York: Springer, 2013), 1–21; S. M. McDougle, K. M. Bond, and J. A. Taylor, "Explicit and Implicit Processes Constitute the Fast and Slow Processes of Sensorimotor Learning," *Journal of Neuroscience* 35 (2015): 9568–79; R. Shadmehr and F. A. Mussa-Ivaldi, "Adaptive Representation of Dynamics during Learning of a Motor Task," *Journal of Neuroscience* 14 (1994): 3208–24.

19. A huge experimental literature exists on the cat and monkey superior colliculus, including how it monitors and makes predictions about head and eye position and predicts how visual images will move across the retina as a consequence. Here I list only a few reviews. M. A. Basso and P. J. May, "Circuits for Action and Cognition: A View from the Superior Colliculus," *Annual Review of Vision Science* 3 (2017): 197–226; D. L. Sparks, "Conceptual Issues Related to the Role of the Superior Colliculus in the Control of Gaze," *Current Opinion in Neurobiology* 9 (1999): 698–707; R. H. Wurtz and J. E. Albano, "Visual-Motor Function of the Primate Superior Colliculus," *Annual Review of Neuroscience* 3 (1980): 189–226.

CHAPTER 4: THE CEREBRAL CORTEX AND CONSCIOUSNESS

1. L. Medina and A. Reiner, "Do Birds Possess Homologues of Mammalian Primary Visual, Somatosensory and Motor Cortices?" *Trends in Neurosciences* 23 (2000): 1–12; R. K. Naumann and G. Laurent, "Function and Evolution of the Reptilian Cerebral Cortex," in *Evolution of Nervous Systems*, ed. J. Kaas (San Diego: Academic Press, 2017), 491–518.

2. R. R. Lemon, *Vanished Worlds: An Introduction to Historical Geology* (Dubuque, IA: William C. Brown, 1993).

3. J. F. Harrison, A. Kaiser, and J. M. VandenBrooks, "Atmospheric Oxygen Level and the Evolution of Insect Body Size," *Proceedings: Biological Science* 277 (2010): 1937–46.

4. R. L. Carroll, "The Origin and Early Radiation of Terrestrial Vertebrates," *Journal of Paleontology* 75 (2001): 1202–13.

5. S. Sahney, M. J. Benton, and H. J. Falcon-Lang, "Rainforest Collapse Triggered Pensylvanian Tetrapod Diversification in Euramerica," *Geology* 38 (2010): 1079–82.

6. R. K. Naumann and G. Laurent, "Function and Evolution of the Reptilian Cerebral Cortex," in *Evolution of Nervous Systems*, ed. J. Kaas (San Diego: Academic Press, 2017), 491–518.

7. M. Leal and B. J. Powell, "Behavioural Flexibility and Problem-Solving in a Tropical Lizard," *Biological Letters* 8 (2012): 28–30; J. D. Manrod, R. Hartdegen, and G. M. Burghardt, "Rapid Solving of a Problem Apparatus by Juvenile Black-Throated Monitor Lizards (*Varanus albigularis albigularis*)," *Animal Cognition* 11 (2008): 267–73; R. T. Mason and M. R. Parker, "Social Behavior and Pheromonal Communication in Reptiles," *Journal of Comparative Physiology A: Neuroethology, Sensory, Neural, and Behavioral Physiology* 196 (2010): 729–49.

8. T. S. Kemp, *The Origin and Evolution of Mammals* (Oxford, UK: Oxford University Press, 2005); A. S. Romer and L. W. Price, "Review of the Pelycosauria," *Geological Society of America, Special Papers* 28 (1940): 1–534.

9. Z. Molnár, J. H. Kaas, J. A. de Carlos, R. F. Hevner, E. Lein, and P. Němec, "Evolution and Development of the Mammalian Cerebral Cortex," *Brain, Behavior, and Evolution* 83 (2014): 126–39.

10. A. B. Butler, "Evolution of the Thalamus: A Morphological and Functional Review," *Thalamus and Related Systems* 4 (2008): 35–58; E. G. Jones, *The Thalamus* (New York: Springer, 1985).

11. P. Senter, "Phylogenetic Taxonomy and the Names of the Major Archosaurian (Reptilia) Clades," *PaleoBios* 25 (2005): 1–7.

12. V. Dinets, "Apparent Coordination and Collaboration in Cooperatively Hunting Crocodilians," *Ethology, Ecology, and Evolution* 27 (2012): 244–50; J. S. Doody, G. M. Burghardt, and V. Dinets, "Breaking the Social-Non-Social Dichotomy: A Role for Reptiles in Vertebrate Social Behavior Research?" *Ethology* 119 (2012): 1–9; L. D. Garrick and J. W. Lang, "Social Signals and Behaviors of Adult Alligators and Crocodiles," *American Zoologist* 17 (1977): 225–39.

13. M. C. Langer, M. D. Ezcurra, J. S. Bittencourt, and F. E. Novas, "The Origin and Early Evolution of Dinosaurs," *Biological Reviews* 85 (2010): 55–110.

14. M. Bronzati, O. W. M. Rauhut, J. S. Bittencourt, and M. C. Langer, "Endocast of the Late Triassic (Carnian) Dinosaur *Saturnalia tupiniquim*: Implications for the Evolution of Brain Tissue in Sauropodomorpha," *Scientific Reports* 7 (2017): 11931; S. W. Rogers, "Allosaurus, Crocodiles, and Birds: Evolutionary Clues from Spiral Computed Tomography of an Endocast," *Anatomical Record* 257 (1999): 162–73.

15. L. M. Witmer and R. C. Ridgely, "New Insights into the Brain, Braincase, and Ear Region of Tyrannosaurs (Dinosauria, Theropoda), with Implications for Sensory Organization and Behavior," *Anatomical Record* 292 (2009): 1266–96.

16. S. L. Brusatte, J. K. O'Connor, and E. D. Jarvis, "The Origin and Diversification of Birds," *Current Biology* 25 (2015): R888–R898; L. M. Chiappe, *Glorified Dinosaurs: The Origin and Early Evolution of Birds* (Hoboken, NJ: John Wiley & Sons, 2007); Z. Zhou, "The Origin and Early Evolution of Birds: Discoveries, Disputes, and Perspectives from Fossil Evidence," *Naturwissenschaften* 91 (2004): 455–71.

17. Q. Ji and S. Ji, "On the Discovery of the Earliest Bird Fossil in China (*Sinosauropteryx*) and the Origin of Birds," *Chinese Geology* 10 (1996): 30–33; M. A. Norell and X. Xu, "Feathered Dinosaurs," *Annual Review of Earth and Planetary Science* 33 (2005): 277–99.

18. H. J. Karten, "Vertebrate Brains and Evolutionary Connectomics: On the Origins of the Mammalian 'Neocortex,'" *Philosophical Transactions of the Royal Society of London, B, Biological Sciences* 370 (2015): 20150060; L. Medina and A. Reiner, "Do Birds Possess Homologues of Mammalian Primary Visual, Somatosensory and Motor Cortices?" *Trends in Neurosciences* 23 (2000): 1–12.

19. G. R. Hunt and R. D. Gray, "Tool Manufacture by New Caledonian Crows: Chipping Away at Human Uniqueness," *Acta Zoologica Sinica (Supplement)* 52 (2006): 622–25; C. Rutz and J. J. St Clair, "The Evolutionary Origins and Ecological Context of Tool Use in New Caledonian Crows," *Behavioral Processes* 89 (2012): 153–65; C. Rutz, S. Sugasawa, J. E. van der Wal, B. C. Klump, and J. J. St Clair, "Tool Bending in New Caledonian Crows," *Royal Society of Open Science* 3 (2016): 160439.

20. S. A. Jelbert, A. H. Taylor, L. G. Cheke, N. S. Clayton, and R. D. Gray, "Using the Aesop's Fable Paradigm to Investigate Causal Understanding of Water Displacement by New Caledonian Crows," *PLoS One* 9 (2014): e92895.

21. D. M. Beck and S. Kastner, "Top-Down and Bottom-Up Mechanisms in Biasing Competition in the Human Brain," *Vision Research* 49 (2009): 1154–65; R. Desimone and J. Duncan, "Neural Mechanisms of Selective Visual Attention," *Annual Review of Neuroscience* 18 (1995): 193–222.

22. The literature on the mosaic of cortical visual areas in primates is immense. Literally thousands of people, including myself, contributed. Here I list just a few informative sources on both monkeys and humans. D. Felleman and D. Van Essen, "Distributed Hierarchical Processing in the Primate Visual Cortex," *Cerebral Cortex* 1 (1991): 1–47; K. Grill-Spector and R. Malach, "The Human Visual Cortex," *Annual Review of Neuroscience* 27 (2004): 649–77; P. Schiller and E. Tehovnik, *Vision and the Visual System* (Oxford, UK: Oxford University Press, 2015); L. G. Ungerleider and J. V. Haxby, "'What' and 'Where' in the Human Brain," *Current Opinion in Neurobiology* 4 (1994): 157–65; D. C. Van Essen, J. W. Lewis, H. A. Drury, N. Hadjikhani, R. B. Tootell, M. Bakircioglu, and M. I. Miller, "Mapping Visual Cortex in Monkeys and Humans Using Surface-Based Atlases," *Vision Research* 41 (2001): 1359–78; L. Wang, R. E. Mruczek, M. J. Arcaro, and S. Kastner, "Probabilistic Maps of Visual Topography in Human Cortex," *Cerebral Cortex* 25 (2015): 3911–31.

23. T. Moore and M. Zirnsak, "Neural Mechanisms of Selective Visual Attention," *Annual Review of Psychology* 68 (2017): 47–72.

24. R. Desimone and J. Duncan, "Neural Mechanisms of Selective Visual Attention," *Annual Review of Neuroscience* 18 (1995): 193–222.

25. G. Alarcon and A. Valentin, *Introduction to Epilepsy* (Cambridge, UK: Cambridge University Press, 2012).

26. R. B. Barlow Jr. and A. J. Fraioli, "Inhibition in the Limulus Lateral Eye in Situ," *Journal of General Physiology* 71 (1978): 699–720; K. Hadeler, "On the Theory of Lateral Inhibition," *Kybernetik* 14 (1974): 161–65.

27. M. Corbetta, G. Patel, and G. L. Shulman, "The Reorienting System of the Human Brain: From Environment to Theory of Mind," *Neuron* 58 (2008): 306–24; K. Igelström and M. S. A. Graziano, "The Inferior Parietal Lobe and Temporoparietal Junction: A Network Perspective," *Neuropsychologia* 105 (2017): 70–83; R. Saxe and L. J. Powell, "It's the Thought That Counts: Specific Brain Regions for One Component of Theory of Mind," *Psychological Science* 17 (2006): 692–9; M. Scolari, K. N. Seidl-Rathkopf, and S. Kastner, "Functions of the Human Frontoparietal Attention Network: Evidence from Neuroimaging," *Current Opinion in Behavioral Sciences* 1 (2015): 32–39; J. L. Vincent, I. Kahn, A. Z. Snyder, M. E. Raichle, and R. L. Buckner, "Evidence for a Frontoparietal Control System Revealed by Intrinsic Functional Connectivity," *Journal of Neurophysiology* 100 (2008): 3328–42; B. T. T. Yeo, F. M. Krienen, J. Sepulcre, M. R. Sabuncu, D. Lashkari, M. Hollinshead, J. L. Roffman, et al., "The Organization of the Human Cerebral Cortex Estimated by Intrinsic Functional Connectivity," *Journal of Neurophysiology* 106 (2011): 1125–65.

28. D. Dennett, *Sweet Dreams: Philosophical Obstacles to a Science of Consciousness (Jean Nicod Lectures)* (Cambridge, MA: MIT Press, 2005).

29. C. L. Colby and M. E. Goldberg, "Space and Attention in Parietal Cortex," *Annual Review of Neuroscience* 22 (1999): 319–49; J. Gottlieb, "From Thought to Action: The Parietal Cortex as a Bridge between Perception, Action, and Cognition," *Neuron* 53 (2007): 9–16; E. J. Tehovnik, M. A. Sommer, I. H. Chou, W. M. Slocum, and P. H. Schiller, "Eye Fields in the Frontal Lobes of Primates," *Brain Research Reviews* 32 (2000): 413–48.

30. C. Eriksen and J. St James, "Visual Attention within and around the Field of Focal Attention: A Zoom Lens Model," *Perception and Psychophysics* 40 (1986): 225–40; M. I. Posner, C. R. Snyder, and B. J. Davidson, "Attention and the Detection of Signals," *Journal of Experimental Psychology* 109 (1980): 160–74.

31. M. Scolari, E. F. Ester, and J. T. Serences, "Feature- and Object-Based Attentional Modulation in the Human Visual System," in *The Oxford Handbook of Attention*, ed. A. C. Norbre and S. Kastner (Oxford, UK: Oxford University Press, 2015), 573–600; S. Treue, "Object- and Feature-Based Attention: Monkey Physiology," in *The Oxford Handbook of Attention*, ed. A. C. Norbre and S. Kastner (Oxford, UK: Oxford University Press, 2015), 601–19.

32. So many scholars have suggested a link between complexity and consciousness that the idea has become a science fiction trope. However, Giulio Tononi has proposed the most systematic, mathematical form of the hypothesis. G. Tononi, *Phi: A Voyage from the Brain to the Soul* (New York: Pantheon, 2012).

33. R. Bshary, W. Wickler, and H. Fricke, "Fish Cognition: A Primate's Eye View," *Animal Cognition* 5 (2002): 1–13.

34. C. Koch, "Consciousness Redux: What Is It Like to Be a Bee?" *Scientific American Mind* 19 (December 2008): 18–19.

35. D. Skrbina, *Panpsychism in the West* (Boston: MIT Press, 2005).

36. B. J. Baars, *A Cognitive Theory of Consciousness* (New York: Cambridge University Press, 1988); S. Dehaene, *Consciousness and the Brain* (New York: Viking Press, 2014).

37. E. Todorov and M. I. Jordan, "Optimal Feedback Control as a Theory of Motor Coordination," *Nature Neuroscience* 5 (2002): 1226–35.

CHAPTER 5: SOCIAL CONSCIOUSNESS

1. M. J. Doherty, *How Children Understand Others' Thoughts and Feelings* (New York: Psychology Press, 2008); U. Frith and C. D. Frith, "Development and Neurophysiology of Mentalizing," *Philosophical Transactions of the Royal Society of London, B, Biological Sciences* 358 (2003): 459–73; D. Premack and G. Woodruff, "Does the Chimpanzee Have a Theory of Mind?" *Behavioral and Brain Sciences* 1 (1978): 515–26.

2. S. Baron-Cohen, *Mindblindness: An Essay on Autism and Theory of Mind* (Cambridge, MA: MIT Press, 1997); C. K. Friesen and A. Kingstone, "The Eyes Have it! Reflexive Orienting Is Triggered by Nonpredictive Gaze," *Psychonomic Bulletin and Review* 5 (1998): 490–95; E. A. Hoffman and J. V. Haxby, "Distinct Representations of Eye Gaze and Identity in the Distributed Human Neural System for Face Perception," *Nature*

Neuroscience 3 (2000): 80–84; L. A. Symons, K. Lee, C. C. Cedrone, and M. Nishimura, "What Are You Looking At? Acuity for Triadic Eye Gaze," *Journal of General Psychology* 131 (2004): 451–69.

3. S. Baron-Cohen, A. M. Leslie, and U. Frith, "Does the Autistic Child Have a 'Theory of Mind'?" *Cognition* 21 (1985): 37–46; H. Wimmer and J. Perner, "Beliefs about Beliefs: Representation and Constraining Function of Wrong Beliefs in Young Children's Understanding of Deception," *Cognition* 13 (1983): 103–28.

4. H. M. Wellman, D. Cross, and J. Watson, "Meta-Analysis of Theory-of-Mind Development: The Truth about False Belief," *Child Development* 72 (2001): 655–84.

5. C. Krupenye, F. Kano, S. Hirata, J. Call, and M. Tomasello, "Great Apes Anticipate That Other Individuals Will Act According to False Beliefs," *Science* 354 (2016): 110–14.

6. N. S. Clayton, "Ways of Thinking: From Crows to Children and Back Again," *Quarterly Journal of Experimental Psychology* 68 (2015): 209–41.

7. J. J. Gibson, *The Ecological Approach to Visual Perception* (Boston: Houghton Mifflin Harcourt, 1979).

8. D. C. Dennett, *The Intentional Stance* (Cambridge, MA: Bradford Books/MIT Press, 1987).

9. C. L. Baker, R. Saxe, and J. B. Tenenbaum, "Action Understanding as Inverse Planning," *Cognition* 113 (2009): 329–49; N. C. Rabinowitz, F. Perbet., F. Song, C. Zhang, S. M. Ali Eslami, and M. Botvinick, "Machine Theory of Mind," *Computer Science arXiv* (2017): 1802.007740; R. Saxe and S. D. Houlihan, "Formalizing Emotion Concepts within a Bayesian Model of Theory of Mind," *Current Opinion in Psychology* 17 (2017): 15–21.

10. D. J. Acheson, *Elementary Fluid Dynamics* (Oxford, UK: Clarendon Press, 1990).

11. A. Guterstam, H. H. Kean, T. W. Webb, F. S. Kean, and M. S. A. Graziano, "An Implicit Model of Other People's Visual Attention as an Invisible, Force-Carrying Beam Projecting from the Eyes," *Proceedings of the National Academy of Sciences USA* (in press).

12. C. G. Gross, "The Fire That Comes from the Eye," *The Neuroscientist* 5 (1999): 58–64.

13. A. Dundes, *The Evil Eye: A Folklore Casebook* (New York: Garland Press, 1981).

14. E. B. Titchner, "The Feeling of Being Stared At," *Science* 8 (1898): 895–7.

15. J. Piaget, *The Child's Conception of the World*, trans. J. Tomlinson and A. Tomlinson (Totowa, NJ: Little, Adams, 1979).

16. G. A. Winer, J. E. Cottrell, V. Gregg, J. S. Fournier, and L. S. Bica, "Fundamentally Misunderstanding Visual Perception: Adults' Belief in Visual Emissions," *American Psychologist* 57 (2002): 417–24; G. A. Winer, J. E. Cottrell, and K. D. Karefilaki, "Images, Words and Questions: Variables That Influence Beliefs about Vision in Children and Adults," *Journal of Experimental Child Psychology* 63 (1996): 499–525.

CHAPTER 6: YODA AND DARTH: HOW CAN WE FIND CONSCIOUSNESS IN THE BRAIN?

1. J. H. Kaas, "The Evolution of Brains from Early Mammals to Humans," *Wiley Interdisciplinary Review of Cognitive Science* 4 (2013): 33–45.

2. J. E. Bogen, "Some Neurophysiologic Aspects of Consciousness," *Seminars in Neurology* 17 (1997): 95–103; G. M. Edelman, J. A. Gally, and B. J. Baars, "Biology of Consciousness,"

Frontiers in Psychology 2 (2011): 4; L. M. Ward, "The Thalamic Dynamic Core Theory of Conscious Experience," *Consciousness and Cognition* 20 (2011): 464–86.

3. E. G. Jones, *The Thalamus* (New York: Springer, 1985).

4. F. C. Crick and C. Koch, "What Is the Function of the Claustrum?" *Philosophical Transactions of the Royal Society of London, B, Biological Sciences* 360 (2005): 1271–79; Y. Goll, G. Atlan, and A. Citri, "Attention: The Claustrum," *Trends in Neurosciences* 38 (2015): 486–95; Z. K. Mohamad, B. Fabrice, B. Abdelrahman, and P. Fabienne, "Electrical Stimulation of a Small Brain Area Reversibly Disrupts Consciousness," *Epilepsy and Behavior* 37 (2014): 32–35.

5. R. Blake, J. Brascamp, and D. J. Heeger, "Can Binocular Rivalry Reveal Neural Correlates of Consciousness?" *Philosophical Transactions of the Royal Society of London, B, Biological Sciences* 369 (2014): 20130211; R. Blake and N. K. Logothetis, "Visual Competition," *Nature Reviews Neuroscience* 3 (2002): 13–21; D. A. Leopold and N. K. Logothetis, "Activity Changes in Early Visual Cortex Reflect Monkeys' Percepts during Binocular Rivalry," *Nature* 379 (1996): 549–53; B. A. Metzger, K. E. Mathewson, E. Tapia, M. Fabiani, G. Gratton, and D. M. Beck, "Regulating the Access to Awareness: Brain Activity Related to Probe-Related and Spontaneous Reversals in Binocular Rivalry," *Journal of Cognitive Neuroscience* 29 (2017): 1089–102; K. Sandberg, B. Bahrami, R. Kanai, G. R. Barnes, M. Overgaard, and G. Rees, "Early Visual Responses Predict Conscious Face Perception within and between Subjects during Binocular Rivalry," *Journal of Cognitive Neuroscience* 25 (2013): 969–85; F. Tong, M. Meng, and R. Blake, "Neural Bases of Binocular Rivalry," *Trends in Cognitive Sciences* 10 (2006): 502–11.

6. R. Blake and N. K. Logothetis, "Visual Competition," *Nature Reviews Neuroscience* 3 (2002): 13–21; D. A. Leopold and N. K. Logothetis, "Activity Changes in Early Visual Cortex Reflect Monkeys' Percepts during Binocular Rivalry," *Nature* 379 (1996): 549–53.

7. R. Blake, J. Brascamp, and D. J. Heeger, "Can Binocular Rivalry Reveal Neural Correlates of Consciousness?" *Philosophical Transactions of the Royal Society of London, B, Biological Sciences* 369 (2014): 20130211; B. A. Metzger, K. E. Mathewson, E. Tapia, M. Fabiani, G. Gratton, and D. M. Beck, "Regulating the Access to Awareness: Brain Activity Related to Probe-Related and Spontaneous Reversals in Binocular Rivalry," *Journal of Cognitive Neuroscience* 29 (2017): 1089–102; K. Sandberg, B. Bahrami, R. Kanai, G. R. Barnes, M. Overgaard, and G. Rees, "Early Visual Responses Predict Conscious Face Perception within and between Subjects during Binocular Rivalry," *Journal of Cognitive Neuroscience* 25 (2013): 969–85; F. Tong, M. Meng, and R. Blake, "Neural Bases of Binocular Rivalry," *Trends in Cognitive Sciences* 10 (2006): 502–11.

8. R. Blake and N. K. Logothetis, "Visual Competition," *Nature Reviews Neuroscience* 3 (2002): 13–21; F. Tong, M. Meng, and R. Blake, "Neural Bases of Binocular Rivalry," *Trends in Cognitive Sciences* 10 (2006): 502–11.

9. K. Wunderlich, K. A. Schneider, and S. Kastner, "Neural Correlates of Binocular Rivalry in the Human Lateral Geniculate Nucleus," *Nature Neuroscience* 8 (2005): 1595–602.

10. M. S. Beauchamp, J. V. Haxby, J. E. Jennings, and E. A. DeYoe "An fMRI Version of the Farnsworth-Munsell 100-Hue Test Reveals Multiple Color-Selective Areas in Human Ventral Occipitotemporal Cortex," *Cerebral Cortex* 9 (1999): 257–63; B. R. Conway, "Color Signals through Dorsal and Ventral Visual Pathways," *Visual Neuroscience* 31 (2014): 197–209.

11. R. Blake and N. K. Logothetis, "Visual Competition," *Nature Reviews Neuroscience* 3 (2002): 13–21; H. H. Li, J. Rankin, J. Rinzel, M. Carrasco, and D. J. Heeger, "Attention Model of Binocular Rivalry," *Proceedings of the National Academy of Sciences USA* 114 (2017): E6192–E6201; F. Tong, M. Meng, and R. Blake, "Neural Bases of Binocular Rivalry," *Trends in Cognitive Sciences* 10 (2006): 502–11.

12. R. Blake and N. K. Logothetis, "Visual Competition," *Nature Reviews Neuroscience* 3 (2002): 13–21; H. H. Li, J. Rankin, J. Rinzel, M. Carrasco, and D. J. Heeger, "Attention Model of Binocular Rivalry," *Proceedings of the National Academy of Sciences USA* 114 (2017): E6192–E6201; F. Tong, M. Meng, and R. Blake, "Neural Bases of Binocular Rivalry," *Trends in Cognitive Sciences* 10 (2006): 502–11.

13. M. S. Beauchamp, J. V. Haxby, J. E. Jennings, and E. A. DeYoe, "An fMRI Version of the Farnsworth-Munsell 100-Hue Test Reveals Multiple Color-Selective Areas in Human Ventral Occipitotemporal Cortex," *Cerebral Cortex* 9 (1999): 257–63.

14. S. E. Bouvier and S. A. Engel, "Behavioral Deficits and Cortical Damage Loci in Cerebral Achromatopsia," *Cerebral Cortex* 16 (2006): 183–91.

15. M. Binder, K. Gociewicz, B. Windey, M. Koculak, K. Finc, J. Nikadon, M. Derda, and A. Cleeremans, "The Levels of Perceptual Processing and the Neural Correlates of Increasing Subjective Visibility," *Consciousness and Cognition* 55 (2017): 106–25; D. Carmel, N. Lavie, and G. Rees, "Conscious Awareness of Flicker in Humans Involves Frontal and Parietal Cortex," *Current Biology* 16 (2006): 907–11; M. S. Christensen, T. Z. Ramsøy, T. E. Lund, K. H. Madsen, and J. B. Rowe, "An fMRI Study of the Neural Correlates of Graded Visual Perception," *Neuroimage* 31 (2006): 1711–25; S. Dehaene and J. P. Changeux, "Experimental and Theoretical Approaches to Conscious Processing," *Neuron* 70 (2011): 200–227; S. Dehaene, L. Naccache, L. Cohen, D. L. Bihan, J. F. Mangin, J. B. Poline, and D. Rivière, "Cerebral Mechanisms of Word Masking and Unconscious Repetition Priming," *Nature Neuroscience* 4 (2001): 752–58.

16. A. Schurger, I. Sarigiannidis, L. Naccache, J. D. Sitt, and S. Dehaene, "Cortical Activity Is More Stable When Sensory Stimuli Are Consciously Perceived," *Proceedings of the National Academy of Sciences USA* 112 (2015): E2083–92.

17. M. Binder, K. Gociewicz, B. Windey, M. Koculak, K. Finc, J. Nikadon, M. Derda, and A. Cleeremans, "The Levels of Perceptual Processing and the Neural Correlates of Increasing Subjective Visibility," *Consciousness and Cognition* 55 (2017): 106–25; D. Carmel, N. Lavie, and G. Rees, "Conscious Awareness of Flicker in Humans Involves Frontal and Parietal Cortex," *Current Biology* 16 (2006): 907–11; M. S. Christensen, T. Z. Ramsøy, T. E. Lund, K. H. Madsen, and J. B. Rowe, "An fMRI Study of the Neural Correlates of Graded Visual Perception," *Neuroimage* 31 (2006): 1711–25; S. Dehaene and J. P. Changeux, "Experimental and Theoretical Approaches to Conscious Processing," *Neuron* 70 (2011): 200–227; S. Dehaene, L. Naccache, L. Cohen, D. L. Bihan, J. F. Mangin, J. B. Poline, and D. Rivière, "Cerebral Mechanisms of Word Masking and Unconscious Repetition Priming," *Nature Neuroscience* 4 (2001): 752–58.

18. T. W. Webb, K. Igelström, A. Schurger, and M. S. A. Graziano, "Cortical Networks Involved in Visual Awareness Independently of Visual Attention," *Proceedings of the National Academy of Sciences USA* 113 (2016): 13923–28.

19. T. J. Buschman and E. K. Miller, "Goal-Direction and Top-Down Control," *Philosophical Transactions of the Royal Society of London, B, Biological Sciences* 369 (2014): 20130471; E. K. Miller and J. D. Cohen, "An Integrative Theory of Prefrontal Cortex Function," *Annual Review of Neurosciences* 24 (2001): 167–202.

20. A. Nieder and E. K. Miller, "Coding of Cognitive Magnitude: Compressed Scaling of Numerical Information in the Primate Prefrontal Cortex," *Neuron* 37 (2003): 149–57.

21. S. C. Rao, G. Rainer, and E. K. Miller, "Integration of What and Where in the Primate Prefrontal Cortex," *Science* 276 (1997): 821–24.

22. D. J. Freedman, M. Riesenhuber, T. Poggio, and E. K. Miller, "Categorical Representation of Visual Stimuli in the Primate Prefrontal Cortex," *Science* 291 (2001): 312–16.

23. R. Levy and P. S. Goldman-Rakic, "Segregation of Working Memory Functions within the Dorsolateral Prefrontal Cortex," *Experimental Brain Research* 133 (2000): 23–32; E. K. Miller, "The 'Working' of Working Memory," *Dialogues in Clinical Neuroscience* 15 (2013): 411–18.

24. B. Odegaard, R. T. Knight, and H. Lau, "Should a Few Null Findings Falsify Prefrontal Theories of Conscious Perception?" *Journal of Neuroscience* 40 (2017): 9593–602.

25. Several broad reviews make the point that a loss or reduction of consciousness is not generally considered a symptom of prefrontal damage. J. Fuster, *The Prefrontal Cortex* (New York: Academic Press, 2015); A. Henri-Bhargava, D. T. Stuss, and M. Freedman, "Clinical Assessment of Prefrontal Lobe Functions," *Continuum, Behavioral Neurology and Psychiatry* 24 (2018): 704–26; T. Shallice and L. Cipolotti, "The Prefrontal Cortex and Neurological Impairments of Active Thought," *Annual Review of Psychology* 69 (2018): 157–80; S. M. Szczepanski and R. T. Knight, "Insights into Human Behavior from Lesions to the Prefrontal Cortex," *Neuron* 83 (2014): 1002–18.

26. Work on parietal-frontal networks has exploded. Here is a bare minimum of sources that cover the main networks mentioned in the book. D. Bzdok, R. Langner, L. Schilbach, O. Jakobs, C. Roski, S. Caspers, A. R. Laird, et al. "Characterization of the Temporo-Parietal Junction by Combining Data-Driven Parcellation, Complementary Connectivity Analyses, and Functional Decoding," *Neuroimage* 81 (2013): 381–92; M. Corbetta, G. Patel, and G. L. Shulman, "The Reorienting System of the Human Brain: From Environment to Theory of Mind," *Neuron* 58 (2008): 306–24; N. U. Dosenbach, D. A. Fair, F. M. Miezin, A. L. Cohen, K. K. Wenger, R. A. Dosenbach, M. D. Fox, A. Z. Snyder, et al., "Distinct Brain Networks for Adaptive and Stable Task Control in Humans," *Proceedings of the National Academy of Sciences USA* 104 (2007): 11073–78; M. D. Fox, M. Corbetta, A. Z. Snyder, J. L. Vincent, and M. E. Raichle, "Spontaneous Neuronal Activity Distinguishes Human Dorsal and Ventral Attention Systems," *Proceedings of the National Academy of Sciences USA* 103 (2006): 10046–51; K. Igelström and M. S. A. Graziano, "The Inferior Parietal Lobe and Temporoparietal Junction: A Network Perspective," *Neuropsychologia* 105 (2017): 70–83; K. M. Igelström, T. W. Webb, and M. S. A. Graziano, "Neural Processes in the Human Temporoparietal Cortex Separated by Localized Independent Component Analysis," *Journal of Neuroscience* 35 (2015): 9432–45; K. M. Igelström, T. W. Webb, and M. S. A. Graziano, "Topographical Organization of Attentional, Social and Memory Processes in the Human Temporoparietal Cortex," *eNeuro* 3 (2016): e0060; R. B. Mars, J. Sallet, U. Schüffelgen, S. Jbabdi, I. Toni, and M. F. Rushworth, "Connectivity-Based Subdivisions of the Human Right Temporoparietal Junction

Area: Evidence for Different Areas Participating in Different Cortical Networks," *Cerebral Cortex* 22 (2012): 1894–903; R. Ptak, "The Frontoparietal Attention Network of the Human Brain: Action, Saliency, and a Priority Map of the Environment," *Neuroscientist* 18 (2012): 502–15; R. Saxe and L. J. Powell, "It's the Thought That Counts: Specific Brain Regions for One Component of Theory of Mind," *Psychological Science* 17 (2006): 692–99; M. Scolari, K. N. Seidl-Rathkopf, and S. Kastner, "Functions of the Human Frontoparietal Attention Network: Evidence from Neuroimaging," *Current Opinion in Behavioral Sciences* 1 (2015): 32–39; J. L. Vincent, I. Kahn, A. Z. Snyder, M. E. Raichle, and R. L. Buckner, "Evidence for a Frontoparietal Control System Revealed by Intrinsic Functional Connectivity," *Journal of Neurophysiology* 100 (2008): 3328–42; B. T. T. Yeo, F. M. Krienen, J. Sepulcre, M. R. Sabuncu, D. Lashkari, M. Hollinshead, J. L. Roffman, et al., "The Organization of the Human Cerebral Cortex Estimated by Intrinsic Functional Connectivity," *Journal of Neurophysiology* 106 (2011): 1125–65.

27. C. Amiez and M. Petrides, "Anatomical Organization of the Eye Fields in the Human and Non-Human Primate Frontal Cortex," *Progress in Neurobiology* 89 (2009): 220–30; L. L. Chen and E. J. Tehovnik, "Cortical Control of Eye and Head Movements: Integration of Movements and Percepts," *European Journal of Neuroscience* 25 (2007): 1253–64; M. H. Grosbras and A. Berthoz, "Parieto-Frontal Networks and Gaze Shifts in Humans: Review of Functional Magnetic Resonance Imaging Data," *Advances in Neurology* 93 (2003): 269–80; E. Lobel, P. Kahane, U. Leonards, M. Grosbras, S. Lehericy, D. Le Bihan, and A. Berthoz, "Localization of Human Frontal Eye Fields: Anatomical and Functional Findings of Functional Magnetic Resonance Imaging and Intracerebral Electrical Stimulation," *Journal of Neurosurgery* 95 (2001): 804–15.

28. R. Caminiti, S. Ferraina, and P. B. Johnson, "The Sources of Visual Information to the Primate Frontal Lobe: A Novel Role for the Superior Parietal Lobule," *Cerebral Cortex* 6 (1996): 319–28; C. S. Konen, R. E. Mruczek, J. L. Montoya, and S. Kastner, "Functional Organization of Human Posterior Parietal Cortex: Grasping- and Reaching-Related Activations Relative to Topographically Organized Cortex," *Journal of Neurophysiology* 109 (2013): 2897–908; L. H. Snyder, A. P. Batista, and R. A. Andersen, "Coding of Intention in the Posterior Parietal Cortex," *Nature* 386 (1997): 167–70.

29. M. G. Di Bono, C. Begliomini, U. Castiello, and M. Zorzi, "Probing the Reaching-Grasping Network in Humans through Multivoxel Pattern Decoding," *Brain and Behavior* 5 (2015): e00412; C. S. Konen, R. E. Mruczek, J. L. Montoya, and S. Kastner, "Functional Organization of Human Posterior Parietal Cortex: Grasping- and Reaching-Related Activations Relative to Topographically Organized Cortex," *Journal of Neurophysiology* 109 (2013): 2897–908; A. Murata, V. Gallese, G. Luppino, M. Kaseda, and H. Sakata, "Selectivity for the Shape, Size, and Orientation of Objects for Grasping in Neurons of Monkey Parietal Area AIP," *Journal of Neurophysiology* 83 (2000): 2580–601; G. Rizzolatti, R. Camarda, L. Fogassi, M. Gentilucci, G. Luppino, and M. Matelli, "Functional Organization of Inferior Area 6 in the Macaque Monkey. II. Area F5 and the Control of Distal Movements," *Experimental Brain Research* 71 (1988): 491–507.

30. D. F. Cooke and M. S. A. Graziano, "Super-Flinchers and Nerves of Steel: Defensive Movements Altered by Chemical Manipulation of a Cortical Motor Area," *Neuron* 43 (2004):

585–93; D. F. Cooke, C. S. R. Taylor, T. Moore, and M. S. A. Graziano, "Complex Movements Evoked by Microstimulation of Area VIP," *Proceedings of the National Academy of Sciences USA* 100 (2003): 6163–68.

31. E. Eger, P. Pinel, S. Dehaene, and A. Kleinschmidt, "Spatially Invariant Coding of Numerical Information in Functionally Defined Subregions of Human Parietal Cortex," *Cerebral Cortex* 25 (2015): 1319–29; A. Nieder and E. K. Miller, "Coding of Cognitive Magnitude: Compressed Scaling of Numerical Information in the Primate Prefrontal Cortex," *Neuron* 37 (2003): 149–57; R. Stanescu-Cosson, P. Pinel, P. F. van De Moortele, D. Le Bihan, L. Cohen, and S. Dehaene, "Understanding Dissociations in Dyscalculia: A Brain Imaging Study of the Impact of Number Size on the Cerebral Networks for Exact and Approximate Calculation," *Brain* 123 (2000): 2240–55.

32. K. M. Igelström, T. W. Webb, and M. S. A. Graziano, "Topographical Organization of Attentional, Social and Memory Processes in the Human Temporoparietal Cortex," *eNeuro* 3 (2016): e0060; Y. T. Kelly, T. W. Webb, J. D. Meier, M. J. Arcaro, and M. S. A. Graziano, "Attributing Awareness to Oneself and to Others," *Proceedings of the National Academy of Sciences USA* 111 (2014): 5012–17; T. W. Webb, K. Igelström, A. Schurger, and M. S. A. Graziano, "Cortical Networks Involved in Visual Awareness Independently of Visual Attention," *Proceedings of the National Academy of Sciences USA* 113 (2016): 13923–28.

33. K. Igelström and M. S. A. Graziano, "The Inferior Parietal Lobe and Temporoparietal Junction: A Network Perspective," *Neuropsychologia* 105 (2017): 70–83; K. M. Igelström, T. W. Webb, and M. S. A. Graziano, "Topographical Organization of Attentional, Social and Memory Processes in the Human Temporoparietal Cortex," *eNeuro* 3 (2016): e0060; R. B. Mars, J. Sallet, U. Schüffelgen, S. Jbabdi, I. Toni, and M. F. S. Rushworth, "Connectivity-Based Subdivisions of the Human Right Temporoparietal Junction Area: Evidence for Different Areas Participating in Different Cortical Networks," *Cerebral Cortex* 22 (2012): 1894–903.

34. R. Saxe and L. J. Powell, "It's the Thought That Counts: Specific Brain Regions for One Component of Theory of Mind," *Psychological Science* 17 (2006): 692–99.

35. Y. T. Kelly, T. W. Webb, J. D. Meier, M. J. Arcaro, and M. S. A. Graziano, "Attributing Awareness to Oneself and to Others," *Proceedings of the National Academy of Sciences USA* 111 (2014): 5012–17; T. W. Webb, K. Igelström, A. Schurger, and M. S. A. Graziano, "Cortical Networks Involved in Visual Awareness Independently of Visual Attention," *Proceedings of the National Academy of Sciences USA* 113 (2016): 13923–28.

36. M. Corbetta, G. Patel, and G. L. Shulman, "The Reorienting System of the Human Brain: From Environment to Theory of Mind," *Neuron* 58 (2008): 306–24; T. Moore and M. Zirnsak, "Neural Mechanisms of Selective Visual Attention," *Annual Review of Psychology* 68 (2017): 47–72; R. Ptak, "The Frontoparietal Attention Network of the Human Brain: Action, Saliency, and a Priority Map of the Environment," *Neuroscientist* 18 (2012): 502–15.

37. K. M. Igelström, T. W. Webb, and M. S. A. Graziano, "Topographical Organization of Attentional, Social and Memory Processes in the Human Temporoparietal Cortex," *eNeuro* 3 (2016): e0060.

38. M. A. Goodale and A. D. Milner, "Separate Visual Pathways for Perception and Action," *Trends in Neurosciences* 15 (1992): 20–25.

39. M. Hurme, M. Koivisto, A. Revonsuo, and H. Railo, "Early Processing in Primary Visual Cortex Is Necessary for Conscious and Unconscious Vision While Late Processing Is Necessary Only for Conscious Vision in Neurologically Healthy Humans," *Neuroimage* 150 (2017): 230–38; F. Tong, "Primary Visual Cortex and Visual Awareness," *Nature Reviews Neuroscience* 4 (2003): 219–29.

40. A. Cowey, "The Blindsight Saga," *Experimental Brain Research* 200 (2010): 3–24; L. Weiskrantz, E. K. Warrington, M. D. Sanders, and J. Marshall, "Visual Capacity in the Hemianopic Field following a Restricted Cortical Ablation," *Brain* 97 (1974): 709–28.

41. T. N. Aflalo and M. S. A. Graziano, "Organization of the Macaque Extrastriate Visual Cortex Re-examined Using the Principle of Spatial Continuity of Function," *Journal of Neurophysiology* 105 (2011): 305–20; D. Felleman and D. Van Essen, "Distributed Hierarchical Processing in the Primate Visual Cortex," *Cerebral Cortex* 1 (1991): 1–47; M. A. Goodale and A. D. Milner, "Separate Visual Pathways for Perception and Action," *Trends in Neurosciences* 15 (1992): 20–25; L. G. Ungerleider and J. V. Haxby, "'What' and 'Where' in the Human Brain," *Current Opinion in Neurobiology* 4 (1994): 157–65.

42. S. Brown and E. Schafer, "An Investigation into the Functions of the Occipital and Temporal Lobes of the Monkey's Brain," *Philosophical Transactions of the Royal Society of London, B, Biological Sciences* 179 (1888): 303–27.

43. P. Broca, "Remarks on the Seat of the Faculty of Articulate Language, Followed by an Observation of Aphemia," *Bulletin de la Societe Anatomique de Paris* 6 (1861): 330–57, trans. G. von Bonin and republished in *Some Papers on the Cerebral Cortex*, ed. G. Von Bonin (Springfield, IL: Charles Thomas Publisher, 1960), 49–72; A. R. Damasio and N. Geschwind, "The Neural Basis of Language," *Annual Review of Neuroscience* 7 (1984): 127–47.

44. J. Zihl, D. von Cramon, and N. Mai, "Selective Disturbance of Movement Vision after Bilateral Brain Damage," *Brain* 106 (1983): 313–40.

45. S. E. Bouvier and S. A. Engel, "Behavioral Deficits and Cortical Damage Loci in Cerebral Achromatopsia," *Cerebral Cortex* 16 (2006): 183–91.

46. C. Gottesmann, "The Neurophysiology of Sleep and Waking: Intracerebral Connections, Functioning and Ascending Influences of the Medulla Oblongata," *Progress in Neurobiology* 59 (1999): 1–54.

47. D. Chalmers, *The Conscious Mind* (Oxford, UK: Oxford University Press); P. Skokowski, "I, Zombie," *Consciousness and Cognition* 11 (2002): 1–9; C. Tandy, "Are You (Almost) a Zombie? Conscious Thoughts about 'Consciousness in the Universe' by Hameroff and Penrose," *Physics of Life Reviews* 11 (2014): 89–90.

48. W. R. Brain, "A Form of Visual Disorientation Resulting from Lesions of the Right Cerebral Hemisphere," *Proceedings of the Royal Society of Medicine* 34 (1941): 771–76; M. Critchley, *The Parietal Lobes* (London: Hafner Press, 1953); G. Vallar, "Extrapersonal Visual Unilateral Spatial Neglect and Its Neuroanatomy," *Neuroimage* 14 (2001): S52–S58.

49. K. M. Heilman and E. Valenstein "Mechanism Underlying Hemispatial Neglect," *Annual Neurology* 5 (1972): 166–70; M. Kinsbourne, "A Model for the Mechanism of Unilateral Neglect of Space," *Transactions of the American Neurological Association* 95 (1970): 143–46; M. M. Mesulam, "A Cortical Network for Directed Attention and Unilateral Neglect," *Annual Neurology* 10 (1981): 309–25; S. M. Szczepanski, C. S. Konen, and

S. Kastner, "Mechanisms of Spatial Attention Control in Frontal and Parietal Cortex," *Journal of Neuroscience* 30 (2010): 148–60.

50. P. Chen and K. M. Goedert, "Clock Drawing in Spatial Neglect: A Comprehensive Analysis of Clock Perimeter, Placement, and Accuracy," *Journal of Neuropsychology* 6 (2012): 270–89.

51. E. Bisiach and C. Luzzatti, "Unilateral Neglect of Representational Space," *Cortex* 14 (1978): 129–33.

52. J. C. Marshall and P. W. Halligan, "Blindsight and Insight in Visuo-Spatial Neglect," *Nature* 336 (1988): 766–67.

53. G. Vallar and D. Perani, "The Anatomy of Unilateral Neglect after Right-Hemisphere Stroke Lesions: A Clinical/CT-Scan Correlation Study in Man," *Neuropsychologia* 24 (1986): 609–22.

54. M. A. Bruno, S. Majerus, M. Boly, A. Vanhaudenhuyse, C. Schnakers, O. Gosseries, P. Boveroux, et al., "Functional Neuroanatomy Underlying the Clinical Subcategorization of Minimally Conscious State Patients," *Journal of Neurology* 259 (2012): 1087–98; S. Laureys, "The Neural Correlate of (Un)Awareness: Lessons from the Vegetative State," *Trends in Cognitive Sciences* 9 (2005): 556–59; S. Laureys, S. Antoine, M. Boly, S. Elincx, M. E. Faymonville, J. Berré, B. Sadzot, et al., "Brain Function in the Vegetative State," *Acta Neurologica Belgica* 102 (2002): 177–85; J. Leon-Carrion, U. Leon-Dominguez, L. Pollonini, M. H. Wu, R. E. Frye, M. R. Dominguez-Morales, and G. Zouridakis, "Synchronization between the Anterior and Posterior Cortex Determines Consciousness Level in Patients with Traumatic Brain Injury," *Brain Research* 1476 (2012): 22–30; D. Roquet, J. R. Foucher, P. Froehlig, F. Renard, J. Pottecher, H. Besancenot, F. Schneider, et al., "Resting-State Networks Distinguish Locked-In from Vegetative State Patients," *Neuroimage: Clinical* 12 (2016): 16–22.

CHAPTER 7: THE HARD PROBLEM AND OTHER PERSPECTIVES ON CONSCIOUSNESS

1. D. Chalmers, "Facing Up to the Problem of Consciousness," *Journal of Consciousness Studies* 2 (1995): 200–219.

2. D. Chalmers, "The Meta-Problem of Consciousness," *The Journal of Consciousness Studies* 25, nos. 9–10 (2018): 6–61.

3. I. A. Newton, "Letter of Mr. Isaac Newton, Professor of the Mathematicks in the University of Cambridge; Containing His New Theory about Light and Colors: Sent by the Author to the Publisher from Cambridge, Febr. 6. 1671/72; In Order to Be Communicated to the Royal Society," *Philosophical Transactions Royal Society* 6 (1671): 3075–87.

4. F. Kammerer, "The Hardest Aspect of the Illusion Problem—And How to Solve It," *Journal of Consciousness Studies* 23 (2016): 124–39; F. Kammerer, "Can You Believe It? Illusionism and the Illusion Meta-Problem," *Philosophical Psychology* 31 (2018): 44–67.

5. S. Blackmore, "Delusions of Consciousness," *Journal of Consciousness Studies* 23 (2016): 52–64; F. Crick, *The Astonishing Hypothesis: The Scientific Search for the Soul* (New York: Scribner, 1995); D. C. Dennett, *Consciousness Explained* (Boston: Back Bay Books, 1991); K. Frankish, "Illusionism as a Theory of Consciousness," *Journal of Consciousness Studies* 23 (2016): 1–39;

B. Hood, *The Self Illusion: How the Social Brain Creates Identity* (Oxford, UK: Oxford University Press, 2012); F. Kammerer, "The Hardest Aspect of the Illusion Problem—And How to Solve It," *Journal of Consciousness Studies* 23 (2016): 124–39.

6. For the argument that consciousness is an illusion serving to make life more rewarding, see: N. Humphrey, *Soul Dust* (Princeton, NJ: Princeton University Press, 2011).

7. S. Glucksberg, *Understanding Figurative Language* (Oxford, UK: Oxford University Press, 2001).

8. V. S. Ramchandran and W. Hirstein, "The Perception of Phantom Limbs," *Brain* 121 (1998): 1603–30; A. Woodhouse, "Phantom Limb Sensation," *Clinical and Experimental Pharmacology and Physiology* 32 (2005): 132–34.

9. V. S. Ramchandran and W. Hirstein, "The Perception of Phantom Limbs," *Brain* 121 (1998): 1603–30.

10. Y. Luo and T. A. Anderson, "Phantom Limb Pain: A Review," *International Anesthesiology Clinics* 54 (2016): 121–39.

11. V. S. Ramchandran and W. Hirstein, "The Perception of Phantom Limbs," *Brain* 121 (1998): 1603–30.

12. G. Vallar and R. Ronchi, "Somatoparaphrenia: A Body Delusion. A Review of the Neuropsychological Literature," *Experimental Brain Research* 192 (2009): 533–51.

13. O. Sacks, *The Man Who Mistook His Wife for a Hat* (New York: Touchstone, 1998), 56.

14. M. Botvinick and J. D. Cohen, "Rubber Hand 'Feels' What Eye Sees," *Nature* 391 (1998): 756.

15. M. S. A. Graziano, *The Spaces between Us: A Story of Neuroscience, Evolution, and Human Nature* (Oxford, UK: Oxford University Press, 2018).

16. O. Blanke and T. Metzinger, "Full-Body Illusions and Minimal Phenomenal Selfhood," *Trends in Cognitive Sciences* 13 (2009): 7–13.

17. B. J. Baars, *A Cognitive Theory of Consciousness* (Cambridge, UK: Cambridge University Press, 1988).

18. S. Dehaene, *Consciousness and the Brain* (New York: Viking Press, 2014).

19. D. Dennett, *Sweet Dreams* (Cambridge, MA: MIT Press, 2005).

20. C. G. Gross, *Brain, Vision, Memory: Tales in the History of Neuroscience* (New York: Bradford Books, 1999).

21. H. Palsson and P. Edwards, *Seven Viking Romances* (Toronto, Canada: Penguin Books, 1985).

22. D. Rosenthal, *Consciousness and Mind* (Oxford, UK: Oxford University Press, 2006); see also R. L. Gennaro, *Consciousness and Self Consciousness: A Defense of the Higher Order Thought Theory of Consciousness* (Philadelphia: John Benjamin's Publishing, 1996); H. Lau and D. Rosenthal, "Empirical Support for Higher-Order Theories of Consciousness," *Trends in Cognitive Sciences* 15 (2011): 365–73.

23. P. Carruthers, "How We Know Our Own Minds: The Relationship between Mindreading and Metacognition," *Behavioral and Brain Sciences* 32 (2009): 121–82; A. Pasquali, B. Timmermans, and A. Cleeremans, "Know Thyself: Metacognitive Networks and Measures of Consciousness," *Cognition* 117 (2010): 182–90; D. M. Rosenthal, "Consciousness, Content, and Metacognitive Judgments," *Consciousness and Cognition* 9 (2000): 203–14.

24. D. D. Hoffman, "The Interface Theory of Perception," in *Object Categorization: Computer and Human Vision Perspectives*, ed. S. Dickinson, M. Tarr, A. Leonardis, and B. Schiele (New York: Cambridge University Press, 2009), 148–65.

25. P. Grimaldi, H. Lau, and M. A. Basso, "There Are Things That We Know That We Know, and There Are Things That We Do Not Know We Do Not Know: Confidence in Decision-Making," *Neuroscience and Biobehavioral Reviews* 55 (2015): 88–97.

26. D. C. Dennett, *Consciousness Explained* (Boston: Back Bay Books, 1991).

27. S. J. Blackmore, "Consciousness in Meme Machines," *Journal of Consciousness Studies* 10 (2003): 19–30.

28. W. James, *Principles of Psychology* (New York: Henry Holt & Co., 1890).

29. A. M. Turing, "On Computable Numbers, with an Application to the Entscheidungsproblem," *Proceedings of the London Mathematical Society* S2-42 (1937): 230–65.

30. C. E. Shannon, "A Mathematical Theory of Communication," *Bell System Technical Journal* 27 (1948): 379–423, 623–56.

31. R. W. Kentridge, C. A. Heywood, and L. Weiskrantz, "Attention without Awareness in Blindsight," *Proceedings: Biological Sciences* 266 (1999): 1805–11; R. W. Kentridge, C. A. Heywood, and L. Weiskrantz, "Spatial Attention Speeds Discrimination without Awareness in Blindsight," *Neuropsychologia* 42 (2004): 831–35.

32. Here I cite only a sample of the large amount of excellent work showing the separation between awareness and attention. Because it is one of the few results directly related to consciousness that can be demonstrated in the lab in a controlled manner, this phenomenon has attracted a great deal of work. U. Ansorge and M. Heumann, "Shifts of Visuospatial Attention to Invisible (Metacontrast-Masked) Singletons: Clues from Reaction Times and Event-Related Potentials," *Advances in Cognitive Psychology* 2 (2006): 61–76; P. Hsieh, J. T. Colas, and N. Kanwisher, "Unconscious Pop-Out: Attentional Capture by Unseen Feature Singletons Only When Top-Down Attention Is Available," *Psychological Science* 22 (2011): 1220–26; J. Ivanoff and R. M. Klein, "Orienting of Attention without Awareness Is Affected by Measurement-Induced Attentional Control Settings," *Journal of Vision* 3 (2003): 32–40; Y. Jiang, P. Costello, F. Fang, M. Huang, and S. He, "A Gender- and Sexual Orientation-Dependent Spatial Attentional Effect of Invisible Images," *Proceedings of the National Academy of Sciences USA* 103 (2006): 17048–52; R. W. Kentridge, T. C. Nijboer, and C. A. Heywood, "Attended but Unseen: Visual Attention Is Not Sufficient for Visual Awareness," *Neuropsychologia* 46 (2008): 864–69; C. Koch and N. Tsuchiya, "Attention and Consciousness: Two Distinct Brain Processes," *Trends in Cognitive Sciences* 11 (2007): 16–22; A. Lambert, N. Naikar, K. McLachlan, and V. Aitken, "A New Component of Visual Orienting: Implicit Effects of Peripheral Information and Subthreshold Cues on Covert Attention," *Journal of Experimental Psychology, Human Perception and Performance* 25 (1999): 321–40; V. A. Lamme, "Separate Neural Definitions of Visual Consciousness and Visual Attention: A Case for Phenomenal Awareness," *Neural Networks* 17 (2004): 861–72; Z. Lin and S. O. Murray, "More Power to the Unconscious: Conscious, but Not Unconscious, Exogenous Attention Requires Location Variation," *Psychological Science* 26 (2015): 221–30; P. A. McCormick, "Orienting Attention without Awareness," *Journal of Experimental Psychology, Human Perception and Performance* 23 (1997): 168–80; L. J. Norman, C. A. Heywood, and R. W. Kentridge, "Object-Based Attention without Awareness," *Psychological Science* 24 (2013): 836–43; Y. Tsushima, Y. Sasaki, and T. Watanabe, "Greater Disruption Due to Failure of Inhibitory Control on an Ambiguous Distractor," *Science* 314 (2006): 1786–88; T. W. Webb, H. H. Kean, and M. S. A. Graziano,

"Effects of Awareness on the Control of Attention," *Journal of Cognitive Neuroscience* 28 (2016): 842–51; G. F. Woodman and S. J. Luck, "Dissociations among Attention, Perception, and Awareness during Object-Substitution Masking," *Psychological Science* 14 (2003): 605–11.

33. Y. Tsushima, Y. Sasaki, and T. Watanabe, "Greater Disruption Due to Failure of Inhibitory Control on an Ambiguous Distractor," *Science* 314 (2006): 1786–88; T. W. Webb, H. H. Kean, and M. S. A. Graziano, "Effects of Awareness on the Control of Attention," *Journal of Cognitive Neuroscience* 28 (2016): 842–51.

34. T. W. Webb, H. H. Kean, and M. S. A. Graziano, "Effects of Awareness on the Control of Attention," *Journal of Cognitive Neuroscience* 28 (2016): 842–51.

35. It would be impossible to cite all the many theories that incorporate the idea that consciousness is related to the integration of information. Here are only a few. B. J. Baars, *A Cognitive Theory of Consciousness* (Cambridge, UK: Cambridge University Press, 1988); A. B. Barrett, "An Integration of Integrated Information Theory with Fundamental Physics," *Frontiers in Psychology* 5 (2014): 63; F. Crick and C. Koch, "Toward a Neurobiological Theory of Consciousness," *Seminars in the Neurosciences* 2 (1990): 263–75; A. Damasio, *Self Comes to Mind: Constructing the Conscious Brain* (New York: Pantheon, 2015); S. Dehaene, *Consciousness and the Brain* (New York: Viking Press, 2014); G. M. Edelman, J. A. Gally, and B. J. Baars, "Biology of Consciousness," *Frontiers in Psychology* 2 (2012): 4; A. K. Engel and W. Singer, "Temporal Binding and the Neural Correlates of Sensory Awareness," *Trends in Cognitive Sciences* 5 (2011): 16–25; S. Grossberg, "The Link between Brain Learning, Attention, and Consciousness," *Consciousness and Cognition* 8 (1999): 1–44; V. A. Lamme, "Towards a True Neural Stance on Consciousness," *Trends in Cognitive Sciences* 10 (2006): 494–501; G. Tononi, M. Boly, M. Massimini, and C. Koch, "Integrated Information Theory: From Consciousness to Its Physical Substrate," *Nature Reviews Neuroscience* 17 (2016): 450–61; C. Von der Malsburg, "The Coherence Definition of Consciousness," in *Cognition, Computation, and Consciousness*, ed. M. Ito, Y. Miyashita, and E. Rolls (Oxford, UK: Oxford University Press, 1997), 193–204; L. M. Ward, "The Thalamic Dynamic Core Theory of Conscious Experience," *Consciousness and Cognition* 20 (2011): 464–86.

36. G. Tononi, *Phi: A Voyage from the Brain to the Soul* (New York: Pantheon, 2012).

37. K. Koffka, *Principles of Gestalt Psychology* (New York: Harcourt, Brace, 1935).

38. B. E. Stein and M. A. Meredith, *The Merging of the Senses* (Cambridge, MA: MIT Press, 1993).

39. Bálint's syndrome, caused by damage to parts of the parietal lobe, may be an example of the disintegration of a unified sensory world when spatial information is compromised. H. Udesen and A. L. Madsen, "Balint's Syndrome—Visual Disorientation," *Ugeskrift for Laeger* 154 (1992): 1492–94.

CHAPTER 8: CONSCIOUS MACHINES

1. M. White, *Isaac Newton: The Last Sorcerer* (New York: Basic Books, 1999).

2. I. Aleksander, *Impossible Minds: My Neurons, My Consciousness* (Singapore: World Scientific, 2014); B. J. Baars and S. Franklin, "Consciousness Is Computational: The LIDA Model of

Global Workspace Theory," *International Journal of Machine Consciousness* 1 (2009): 23–32; A. Chella and R. Manzotti, "Machine Consciousness: A Manifesto for Robotics," *International Journal of Machine Consciousness* 1 (2009): 33–51; L. A. Coward and R. Sun, "Criteria for an Effective Theory of Consciousness and Some Preliminary Attempts," *Consciousness and Cognition* 13 (2004): 268–301; S. Franklin, "IDA: A Conscious Artefact," in *Machine Consciousness*, ed. O. Holland (Exeter, UK: Imprint Academic, 2003); P. Haikonen, *Consciousness and Robot Sentience* (Singapore: World Scientific, 2012); O. Holland and R. Goodman, "Robots with Internal Models: A Route to Machine Consciousness?" *Journal of Consciousness Studies* 10 (2003): 77–109; N. Marupaka, L. Lyer, and A. Minai, "Connectivity and Thought: The Influence of Semantic Network Structure in a Neurodynamical Model of Thinking," *Neural Networks* 32 (2012): 147–58; D. Rudrauf, D. Bennequin, I. Granic, G. Landini, K. Friston, and K. Williford, "A Mathematical Model of Embodied Consciousness," *Journal of Theoretical Biology* 428 (2017): 106–31; M. Shanahan, "A Cognitive Architecture That Combines Internal Simulation with a Global Workspace," *Consciousness and Cognition* 15 (2006): 443–49.

3. A. M. Turing, "Computing Machinery and Intelligence," *Mind* 59 (1950): 433–60.

4. S. Baron-Cohen, A. M. Leslie, and U. Frith, "Does the Autistic Child Have a 'Theory of Mind?'" *Cognition* 21 (1985): 37–46; H. Wimmer and J. Perner, "Beliefs about Beliefs: Representation and Constraining Function of Wrong Beliefs in Young Children's Understanding of Deception," *Cognition* 13 (1983): 103–28.

5. N. S. Clayton, "Ways of Thinking: From Crows to Children and Back Again," *Quarterly Journal of Experimental Psychology* 68 (2015): 209–41; C. Krupenye, F. Kano, S. Hirata, J. Call, and M. Tomasello, "Great Apes Anticipate That Other Individuals Will Act According to False Beliefs," *Science* 354 (2016): 110–14; H. M. Wellman, D. Cross, and J. Watson, "Meta-Analysis of Theory-of-Mind Development: The Truth about False Belief," *Child Development* 72 (2001): 655–84.

6. M. A. Lebedev and M. A. Nicolelis, "Brain-Machine Interfaces: From Basic Science to Neuroprostheses and Neurorehabilitation," *Physiological Review* 97 (2017): 767–837.

7. J. V. Haxby, M. I. Gobbini, M. L. Furey, A. Ishai, J. L. Schouten, and P. Pietrini, "Distributed and Overlapping Representations of Faces and Objects in Ventral Temporal Cortex," *Science* 293 (2001): 2425–30.

8. Attention has been built into artificial devices so often that I can cite only a very partial list. H. Adeli, F. Vitu, and G. F. Zelinsky, "A Model of the Superior Colliculus Predicts Fixation Locations during Scene Viewing and Visual Search," *Journal of Neuroscience* 37 (2017): 1453–67; A. Borji and L. Itti, "State-of-the-Art in Visual Attention Modeling," *IEEE Transactions on Pattern Analysis and Machine Intelligence* 35 (2013): 185–207; G. Deco and E. T. Rolls, "A Neurodynamical Cortical Model of Visual Attention and Invariant Object Recognition," *Vision Research* 44 (2004): 621–42; Y. Fang, C. Zhang, J. Li, J. Lei, M. Perreira Da Silva, and P. Le Callet, "Visual Attention Modeling for Stereoscopic Video: A Benchmark and Computational Model," *IEEE Transactions on Image Processing* 26 (2017): 4684–96; S. Goferman, L. Zelnikmanor, and A. Tal, "Context-Aware Saliency Detection," *IEEE Transactions on Pattern Analysis and Machine Intelligence* 34 (2012): 1915–26; C. Guo and L. Zhang, "A Novel Multi-Resolution Spatiotemporal Saliency Detection Model and Its Applications in Image and Video Compression," *IEEE Transactions on Image Processing* 19 (2010): 185–98;

L. Itti, C. Koch, and E. Niebur, "A Model of Saliency-Based Visual Attention for Rapid Scene Analysis," *IEEE Transactions on Pattern Analysis and Machine Intelligence* 20 (1988): 1254–59; O. Le Meur, P. Le Callet, and D. Barba, "A Coherent Computational Approach to Model the Bottom-Up Visual Attention," *IEEE Transactions on Pattern Analysis and Machine Intelligence* 28 (2006): 802–17; R. J. Lin and W. S. Lin, "A Computational Visual Saliency Model Based on Statistics and Machine Learning," *Journal of Vision* 14 (2014): 1; T. Miconi and R. VanRullen, "A Feedback Model of Attention Explains the Diverse Effects of Attention on Neural Firing Rates and Receptive Field Structure," *PLoS Computational Biology* 12 (2016): e1004770; J. H. Reynolds and D. J. Heeger, "The Normalization Model of Attention," *Neuron* 61 (2009): 168–85; P. Schwedhelm, B. S. Krishna, and S. Treue, "An Extended Normalization Model of Attention Accounts for Feature-Based Attentional Enhancement of Both Response and Coherence Gain," *PLoS Computational Biology* 12 (2016): e1005225; M. A. Schwemmer, S. F. Feng, P. J. Holmes, J. Gottlieb, and J. D. Cohen, "A Multi-Area Stochastic Model for a Covert Visual Search Task," *PLoS One* 10 (2015): e0136097; S. Vossel, C. Mathys, K. E. Stephan, and K. J. Friston, "Cortical Coupling Reflects Bayesian Belief Updating in the Deployment of Spatial Attention," *Journal of Neuroscience* 35 (2015): 11532–42; A. L. White, M. Rolfs, and M. Carrasco, "Stimulus Competition Mediates the Joint Effects of Spatial and Feature-Based Attention," *Journal of Vision* 15 (2015): doi 10.1167/15.14.7; P. Zhang, T. Zhuo, W. Huang, K. Chen, and M. Kankanhalli, "Online Object Tracking Based on CNN with Spatial-Temporal Saliency Guided Sampling," *Neurocomputing* 257 (2017): 115–27.

9. E. van den Boogaard, J. Treur, and M. Turpijn, "A Neurologically Inspired Neural Network Model for Graziano's Attention Schema Theory for Consciousness," *International Work Conference on the Interplay between Natural and Artificial Computation: Natural and Artificial Computation for Biomedicine and Neuroscience* Part 1 (2017): 10–21.

10. M. M. Chun, J. D. Golomb, and N. B. Turk-Browne, "A Taxonomy of External and Internal Attention," *Annual Review of Psychology* 62 (2011): 73–101.

11. J. Ledoux, *The Emotional Brain: The Mysterious Underpinnings of Emotional Life* (New York: Simon & Schuster, 1998).

12. W. R. Hess, *Functional Organization of the Diencephalons* (New York: Grune and Stratton, 1957).

13. B. G. Hoebel, "Neuroscience and Appetitive Behavior Research: 25 Years," *Appetite* 29 (1997): 119–33; T. V. Sewards and M. A. Sewards, "Representations of Motivational Drives in Mesial Cortex, Medial Thalamus, Hypothalamus and Midbrain," *Brain Research Bulletin* 61 (2003): 25–49; A. Venkatraman, B. L. Edlow, and M. H. Immordino-Yang, "The Brainstem in Emotion: A Review," *Frontiers in Neuroanatomy* 11 (2017): 15.

14. J. LeDoux, "The Amygdala," *Current Biology* 17 (2007): R868–R874; P. J. Walen and E. A. Phelps, *The Human Amygdala* (New York: Guilford Press, 2009).

15. E. T. Rolls and F. Grabenhorst, "The Orbitofrontal Cortex and Beyond: From Affect to Decision-Making," *Progress in Neurobiology* 86 (2008): 216–44.

16. M. Tamietto and B. de Gelder, "Neural Bases of the Non-Conscious Perception of Emotional Signals," *Nature Reviews Neuroscience* 11 (2010): 697–709; P. Winkielman and K. C. Berridge, "Unconscious Emotion," *Current Directions in Psychological Science* 13 (2004): 120–23.

17. J. E. LeDoux and R. Brown, "A Higher-Order Theory of Emotional Consciousness," *Proceedings of the National Academy of Sciences, USA* 114 (2017): E2016–E2025.

18. W. Cannon, "The James-Lange Theory of Emotions: A Critical Examination and an Alternative Theory," *The American Journal of Psychology* 39 (1927): 106–24.

19. D. G. Dutton and A. P. Aaron, "Some Evidence for Heightened Sexual Attraction under Conditions of High Anxiety," *Journal of Personality and Social Psychology* 30 (1974): 510–17.

20. M. E. Moran, "The da Vinci Robot," *Journal of Endourology* 20 (2006): 986–90.

21. I. Asimov, *The Bicentennial Man* (New York: Ballantine Books, 1976).

22. P. K. Dick, *Do Androids Dream of Electric Sheep?* (New York: Doubleday, 1968).

23. D. Levy, "The Ethical Treatment of Artificially Conscious Robots," *International Journal of Social Robotics* 1 (1929): 209–16.

24. B. Hood, *The Self Illusion: How the Social Brain Creates Identity* (Oxford, UK: Oxford University Press, 2012); F. Podschwadek, "Do Androids Dream of Normative Endorsement? On the Fallibility of Artificial Moral Agents," *Artificial Intelligence and Law* 25 (2017): 325–39; J. Sullins, "Artificial Phronesis and the Social Robot," *Frontiers in Artificial Intelligence and Applications* 290 (2016): 37–39.

CHAPTER 9: UPLOADING MINDS

1. R. Blackford and D. Broderick, eds., *Intelligence Unbound: The Future of Uploads and Machine Minds* (Hoboken, NJ: Wiley Blackwell, 2014); C. Eliasmith, T. C. Stewart, X. Choo, T. Bekolay, T. DeWolf, Y. Tang, and D. Rasmussen, "A Large-Scale Model of the Functioning Brain," *Science* 338 (2012): 1202–05; D. Eth, J.-C. Foust, and B. Whale, "The Prospects of Whole Brain Emulation within the Next Half-Century," *Journal of Artificial General Intelligence* 4 (2013): 130–52; R. A. Koene, "Feasible Mind Uploading," in *Intelligence Unbound: The Future of Uploaded and Machine Minds*, ed. R. Blackford and D. Broderick (Hoboken, NJ: Wiley-Blackwell, 2014); R. Kurzweil, *The Singularity Is Near: When Humans Transcend Biology* (New York: Penguin Books, 2006); H. Markram, E. Muller, S. Ramaswamy, M. Reimann, M. Abdellah, C. A. Sanchez, A. Ailamaki, et al., "Reconstruction and Simulation of Neocortical Microcircuitry," *Cell* 163 (2015): 456–92; H. Moravec, *Mind Children: The Future of Robot and Human Intelligence* (Cambridge, MA: Harvard University Press, 1988).

2. S. Herculano-Houzel, "The Human Brain in Numbers: A Linearly Scaled-Up Primate Brain," *Frontiers in Human Neuroscience* 3 (2009). doi: 10.3389/neuro.09.031.2009

3. C. S. Sherrington, "Santiago Ramón y Cajal 1852–1934," *Biographical Memoirs of Fellows of the Royal Society* 1 (1935): 424–41.

4. S. R. Cajal, J. DeFelipe, and E. G. Jones, *Cajal on the Cerebral Cortex: An Annotated Translation of the Complete Writings* (Oxford, UK: Oxford University Press, 1988).

5. D. E. Rumelhart and J. McClelland, *Parallel Distributed Processing: Explorations in the Microstructure of Cognition* (Cambridge, MA: MIT Press, 1986); J. Schmidhuber, "Deep Learning in Neural Networks: An Overview," *Neural Networks* 61 (2015): 85–117.

6. D. M. Barch, "Resting-State Functional Connectivity in the Human Connectome Project: Current Status and Relevance to Understanding Psychopathology," *Harvard Review*

of Psychiatry 25 (2017): 209–17; D. D. Bock, W. C. Lee, A. M. Kerlin, M. L. Andermann, G. Hood, A. W. Wetzel, S. Yurgenson, et al., "Network Anatomy and *In Vivo* Physiology of Visual Cortical Neurons," *Nature* 471 (2011): 177–82; G. Gong, Y. He, L. Concha, C. Lebel, D. W. Gross, A. C. Evans, and C. Beaulieu, "Mapping Anatomical Connectivity Patterns of Human Cerebral Cortex Using *In Vivo* Diffusion Tensor Imaging Tractography," *Cerebral Cortex* 19 (2009): 524–36; P. Hagmann, L. Cammoun, X. Gigandet, R. Meuli, C. J. Honey, V. J. Wedeen, and O. Sporns, "Mapping the Structural Core of Human Cerebral Cortex," *PLoS Biol* 6 (2008): e159; P. Hagmann, M. Kurant, X. Gigandet, P. Thiran, V. J. Wedeen, R. Meuli, and J.-P. Thiran, "Mapping Human Whole-Brain Structural Networks with Diffusion MRI," *PLoS One* 2 (2007): e597; M. Helmstaedter, K. L. Briggman, S. C. Turaga, V. Jain, H. S. Seung, and W. Denk, "Connectomic Reconstruction of the Inner Plexiform Layer in the Mouse Retina," *Nature* 500 (2013): 168–74; O. Sporns, G. Tononi, and R. Kötter, "The Human Connectome: A Structural Description of the Human Brain," *PLoS Computational Biology* 1 (2005): e42; L. R. Varshney, B. L. Chen, E. Paniagua, D. H. Hall, and D. B. Chklovskii, "Structural Properties of the *Caenorhabditis elegans* Neuronal Network," *PLoS Computational Biology* 7 (2011): e1001066; Z. Zheng, J. S. Lauritzen, E. Perlman, C. G. Robinson, M. Nichols, D. Milkie, O. Torrens, et al., "A Complete Electron Microscopy Volume of the Brain of Adult *Drosophila melanogaster*," *Cell* 174 (2018): 730–43.

7. L. R. Varshney, B. L. Chen, E. Paniagua, D. H. Hall, and D. B. Chklovskii, "Structural Properties of the *Caenorhabditis elegans* Neuronal Network," *PLoS Computational Biology* 7 (2011): e1001066; Z. Zheng, J. S. Lauritzen, E. Perlman, C. G. Robinson, M. Nichols, D. Milkie, O. Torrens, et al., "A Complete Electron Microscopy Volume of the Brain of Adult *Drosophila melanogaster*," *Cell* 174 (2018): 730–43.

8. D. D. Bock, W. C. Lee, A. M. Kerlin, M. L. Andermann, G. Hood, A. W. Wetzel, S. Yurgenson, et al., "Network Anatomy and *In Vivo* Physiology of Visual Cortical Neurons," *Nature* 471 (2011): 177–82

9. D. M. Barch, "Resting-State Functional Connectivity in the Human Connectome Project: Current Status and Relevance to Understanding Psychopathology," *Harvard Review of Psychiatry* 25 (2017): 209–17; G. Gong, Y. He, L. Concha, C. Lebel, D. W. Gross, A. C. Evans, and C. Beaulieu, "Mapping Anatomical Connectivity Patterns of Human Cerebral Cortex Using *In Vivo* Diffusion Tensor Imaging Tractography," *Cerebral Cortex* 19 (2009): 524–36; P. Hagmann, L. Cammoun, X. Gigandet, R. Meuli, C. J. Honey, V. J. Wedeen, and O. Sporns, "Mapping the Structural Core of Human Cerebral Cortex," *PLoS Biol* 6 (2008): e159; P. Hagmann, M. Kurant, X. Gigandet, P. Thiran, V. J. Wedeen, R. Meuli, and J.-P. Thiran, "Mapping Human Whole-Brain Structural Networks with Diffusion MRI," *PLoS One* 2 (2007): e597; O. Sporns, G. Tononi, and R. Kötter, "The Human Connectome: A Structural Description of the Human Brain," *PLoS Computational Biology* 1 (2005): e42.

10. S. Herculano-Houzel, "The Human Brain in Numbers: A Linearly Scaled-Up Primate Brain," *Frontiers in Human Neuroscience* 3 (2009). doi: 10.3389/neuro.09.031.2009

11. N. A. O'Rourke, N. C. Weiler, K. D. Micheva, and S. J. Smith, "Deep Molecular Diversity of Mammalian Synapses: Why It Matters and How to Measure It," *Nature Reviews*

Neuroscience 13 (2012): 365–79; V. Pickel and M. Segal, *The Synapse: Structure and Function* (New York: Academic Press, 2014).

12. B. A. Barres, B. Stevens, and M. R. Freeman, *Glia* (Cold Spring Harbor, NY: Cold Spring Harbor Laboratory Press, 2014).

13. A. Einstein, *The Collected Papers of Albert Einstein: Vol. 7: The Berlin Years: Writings, 1918–1921*, trans. A. Engel (Princeton, NJ: Princeton University Press, 2002).

14. B. P. Abbott et al. (LIGO Scientific Collaboration and Virgo Collaboration), "Observation of Gravitational Waves from a Binary Black Hole Merger," *Physical Review Letters* 116 (2016): 061102.

15. M. L. Cappuccio, "Mind-Upload. The Ultimate Challenge to the Embodied Mind Theory," *Phenomenology and the Cognitive Sciences* 16 (2017): 425–48; M. Wheeler, "Cognition in Context: Phenomenology, Situated Robotics and the Frame Problem," *International Journal of Philosophical Studies* 16 (2008): 323–49.

16. O. Blanke and T. Metzinger, "Full-Body Illusions and Minimal Phenomenal Selfhood," *Trends in Cognitive Sciences* 13 (2009): 7–13; M. S. A. Graziano and M. M. Botvinick, "How the Brain Represents the Body: Insights from Neurophysiology and Psychology," in *Common Mechanisms in Perception and Action: Attention and Performance XIX*, ed. W. Prinz and B. Hommel (Oxford, UK: Oxford University Press, 2002), 136–57; C. Lopez, "Making Sense of the Body: The Role of Vestibular Signals," *Multisensory Research* 28 (2015): 525–57; A. Serino, A. Alsmith, M. Costantini, A. Mandrigin, A. Tajadura-Jimenez, and C. Lopez, "Bodily Ownership and Self-Location: Components of Bodily Self-Consciousness," *Consciousness and Cognition* 22 (2013): 1239–52.

17. An article on the simulated arm was never published. For a review of my work on movement control, see: M. S. A. Graziano, *The Intelligent Movement Machine* (Oxford, UK: Oxford University Press, 2008).

18. T. D. Bancroft, "Ethical Aspects of Computational Neuroscience," *Neuroethics* 6 (2013): 415–18; P. Eckersley and A. Sandberg, "Is Brain Emulation Dangerous?" *Journal of Artificial General Intelligence* 4 (2013): 170–94; K. Muzyka, "The Outline of Personhood Law Regarding Artificial Intelligences and Emulated Human Entities," *Journal of Artificial General Intelligence* 4 (2013): 164–69.

19. D. Sheils, "Toward a Unified Theory of Ancestor Worship: A Cross-Cultural Study," *Social Forces* 54 (1975): 427–40.

20. B. B. Powell, *Writing: Theory and History of the Technology of Civilization* (Oxford, UK: Blackwell Press, 2009).

21. J. P. Mallory and D. Q. Adams, *The Oxford Introduction to Proto-Indo-European and the Proto-Indo-European World* (Oxford, UK: Oxford University Press, 2006).

22. G. Santayana, *Reason in Common Sense* (New York: Dover, 1980).

APPENDIX: HOW TO BUILD VISUAL CONSCIOUSNESS

1. R. Klette, *Concise Computer Vision* (New York: Springer, 2014); L. G. Shapiro and G. C. Stockman, *Computer Vision* (Upper Saddle River, NJ: Prentice Hall, 2001); M. Sonka, V. Hlavac,

and R. Boyle, *Image Processing, Analysis, and Machine Vision* (Stamford, CT: Cengage Learning, 2008).

2. P. M. Merikle, D. Smilek, and J. D. Eastwood, "Perception without Awareness: Perspectives from Cognitive Psychology," *Cognition* 79 (2001): 115–34; R. Szczepanowski and L. Pessoa, "Fear Perception: Can Objective and Subjective Awareness Measures Be Dissociated?" *Journal of Vision* 10 (2007): 1–17.

3. M. Tegmark, "Consciousness as a State of Matter," *arXiv* (2014): 1401.1219; G. Tononi, M. Boly, M. Massimini, and C. Koch, "Integrated Information Theory: From Consciousness to Its Physical Substrate," *Nature Reviews Neuroscience* 17 (2016): 450–61.

4. M. S. A. Graziano and M. M. Botvinick, "How the Brain Represents the Body: Insights from Neurophysiology and Psychology," in *Common Mechanisms in Perception and Action: Attention and Performance XIX*, ed. W. Prinz and B. Hommel (Oxford, UK: Oxford University Press, 2002), 136–57; N. Holmes and C. Spence, "The Body Schema and the Multisensory Representation(s) of Personal Space," *Cognitive Processing* 5 (2004): 94–105; F. de Vignemont, *Mind the Body: An Exploration of Bodily Self-Awareness* (Oxford, UK: Oxford University Press, 2018).

5. S. Bluck and T. Habermas, "The Life Story Schema," *Motivation and Emotion* 24 (2000): 121–47; M. A. Conway and C. W. Pleydell-Pearce, "The Construction of Autobiographical Memories in the Self-Memory System," *Psychological Review* 107 (2000): 261–88; M. A. Conway, J. A. Singer, and A. Tagini, "The Self and Autobiographical Memory: Correspondence and Coherence," *Social Cognition* 22 (2004): 491–529.

6. O. Blanke, "Multisensory Brain Mechanisms of Bodily Self-Consciousness," *Nature Reviews Neuroscience* 13 (2012): 556–71; O. Blanke and T. Metzinger, "Full-Body Illusions and Minimal Phenomenal Selfhood," *Trends in Cognitive Sciences* 13 (2009): 7–13; C. Preston, B. J. Kuper-Smith, and H. H. Ehrsson, "Owning the Body in the Mirror: The Effect of Visual Perspective and Mirror View on the Full-Body Illusion," *Scientific Reports* 5 (2015): 18345.

7. M. S. Gazzaniga, *The Bisected Brain* (New York: Appleton Century Crofts, 1970); R. E. Nisbett and T. D. Wilson, "Telling More Than We Can Know—Verbal Reports on Mental Processes," *Psychological Review* 84 (1977): 231–59.

8. D. M. Beck and S. Kastner, "Top-Down and Bottom-Up Mechanisms in Biasing Competition in the Human Brain," *Vision Research* 49 (2009): 1154–65; R. Desimone and J. Duncan, "Neural Mechanisms of Selective Visual Attention," *Annual Review of Neuroscience* 18 (1995): 193–222.

9. G. Deco and E. T. Rolls, "Neurodynamics of Biased Competition and Cooperation for Attention: A Model with Spiking Neurons," *Journal of Neurophysiology* 94 (2005): 295–313; L. Layon and S. L. Denham, "A Biased Competition Computational Model of Spatial and Object-Based Attention Mediating Active Visual Search," *Neurocomputing* 58 (2004): 655–62; J. Reynolds and D. Heeger, "The Normalization Model of Attention," *Neuron* 61 (2009): 168–85; J. K. Tsotsos, *A Computational Perspective on Visual Attention* (Cambridge, MA: MIT Press, 2011).

INDEX

Note: Page numbers in italics indicate figures.

James, William, 109–10, 111, 131–32
James-Lange theory, 131–32
Joyce, James, *Ulysses*, 4
Jurassic period, 29

Kammerer, François, 94–95
Kastner, Sabine, 67
Kentridge, Robert, 111–12
knowledge
  model-based, 3, 50, 58–64, 92, 135–36
    (*see also* attention schema; internal
    models; self models)
  preservation of, 156–61
kraken, 106–7
Kubrick, Stanley, 133

Lange, Carl, 131–32
lateral geniculate nucleus, *31*, 32
lateral inhibition, 10–11, 33–34
LeDoux, Joseph, 130–31
Leonardo da Vinci, 133
Liaoning fossil bed, 29
local inhibition, 33
location information, 114–15
Loebner Prize competition, 122
Lucas, George, 133

machine consciousness, 7. *See also* conscious
  machines
  visual consciousness and, 133, *168*, *171*,
  174
mammals, 25, *45*. *See also specific mammals*
  cerebral cortex in, 26, 27, 30
  consciousness and, 45
The Matrix, 135
memory, 5, 6, 150–51
mental models, 121
Mesozoic era, 29
meta-information, 107. *See also* higher-order
  thought theory
metaphors, 96–101
metaphysicality, 96
meta-problem, 91–96
Metzinger, Thomas, 104–5
Miller, Earl, 76–77
Milner, David, 81–82
mind, as information, 85

mind uploading, 138–66
  afterlife and, 151, 155–56
  capacity limits, 150–51
  embodiment and, 148–49
  human sociality and, 164–65
  identity and, 152–54
  immersive worlds and, 149–50
  impracticality of, 150
  memory and, 150–51
  mind melding and, 164–65
  philosophical and cultural questions,
  152–61
  potential pitfalls of, 150–52
  preservation of knowledge and,
  152–61
  resources for, 151–52
  rights and, 151
  scanning and, 139–48
  simulation and, 146–48
  space travel and, 161–64
model-based knowledge, 3, 50, 58–64, 135–
  36. *See also* attention schema; internal
  models; self models
  vs. superficial knowledge, 92
model of attention. *See* attention schema
mollusks, *9*, 12–13
monkeys, 20
  binocular rivalry and, 68
  visual system in, 31–33, *31*
motion blindness, brain damage and, 86
movement control, 34
MRIs (magnetic resonance imaging), 124,
  142–44

National Institutes of Health (NIH),
  142–43
Nelson, Horatio, Lord, 101, 104
nerves, 9–10
nervous system, 7, 8–15, *45*
neural networks, 140–44, 147
neuron doctrine, 140
neurons, 9, 32–33, 77, 139–45, 147
  binocular rivalry and, 68
  in frogs, 18–19
  measuring activity of, 18–19
neuroscience, 2, 66–67
Newton, Isaac, 94, 118–19